THE INTEGRATED PRACTITIONER

Integrating Everything

BOOK 4 OF *THE INTEGRATED PRACTITIONER* SERIES

JUSTIN AMERY

Radcliffe Publishing
London • New York

Radcliffe Publishing Ltd
St Mark's House
Shepherdess Walk
London N1 7LH
United Kingdom

www.radcliffehealth.com

British Library Cataloguing in Publication Data

A catalogue record for this book is available from the British Library.

ISBN-13: 978 184619 775 8
Volume set ISBN-13: 978 184619 950 9

The paper used for the text pages of this book is FSC® certified. FSC (The Forest Stewardship Council®) is an international network to promote responsible management of the world's forests.

Typeset and designed by Darkriver Design, Auckland, New Zealand
Printed and bound by Hobbs the Printers, Totton, Hants, UK

Contents

These books are dedicated to my Dad, Tony Amery, who was a wonderful doctor and who is still my inspiration.

About the author

I am a full-time practising family practitioner and children's palliative care specialist doctor working in the UK. I have also spent some years working in Uganda and other sub-Saharan African countries.

I enjoy teaching, writing and mentoring. I am a medical student tutor at the University of Oxford, a trainer in general practice, and I have designed and set up children's palliative care courses for health professionals in the UK and Africa. I have worked with 'failing practices' to help them turn round; and also with health professionals who are struggling (as we all do from time to time).

I have always had an interest in philosophy and spirituality, and have studied this at postgraduate level. I have carried out some research into education and training of health professionals around the world and I continue to explore that interest.

I have previously written two books: *Children's Palliative Care in Africa* (Oxford: Oxford University Press, 2009) and the Association for Children's Palliative Care (ACT) *Handbook of Children's Palliative Care for GPs* (Bristol: ACT, 2011). I particularly enjoy reading and writing poetry.

At heart, though, I am a practitioner and a generalist. What is more, as you can probably see, I am rather a jack of all trades, and a master of none.

I have been motivated to write this book as I am hoping to explore practical ways of practising health that help us all, patients and practitioners alike, to become a little more healthy, and a little more whole.

Acknowledgements

These books have been brewing up over many years and so there have been very, very many influences upon them. There are far too many people to mention and thank without risking leaving someone out, so I shall just mention those who have been immediately involved.

Firstly, thank you to those very kind and patient people who helped review the drafts and gave such helpful feedback: Maria Ward, Penny Thompson, Meriel Lynch, Tom Nicholson-Lailey, Peter Burke, Penny Moore, Susan McCrae, Caitlin Chasser, Louise Rutter, Polly Steele, Rachel Samson, Laura Ingle and Maddy Podichetty.

I would also particularly like to mention Chris Smith, who not only gave very useful feedback on these books, but who also helped me to develop a lot of the ideas in them through his leadership of the Oxford Advanced Consultation Skills Course that I help him with, and over a few pints in the pub as well.

Thanks as well to Gillian Nineham of Radcliffe Publishing, who was brave (or daft) enough to put her faith in these rather unconventional offerings; suggest numerous areas for improvement and offer tremendous support and encouragement in their publication. Thanks also to Jamie Etherington and Camille Lowe for all their help in putting them together.

I would like to thank my colleagues at Bury Knowle Health Centre in Oxford, Helen House Hospice in Oxford, Hospice Africa in Kampala, Uganda, and Keech Hospice in Luton. They have all shown utmost patience and perseverance as I have led them on various merry dances, contortions and deviations in the name of 'good ideas', rarely reminding me of the 99% which failed, and always supportive of the 1% that, miraculously, did.

Of course I can't forget Karen Bateman (the doctor) and Karen Amery (the missus) who has been a continuous and never-ending source of sound advice, support and wisdom.

Finally, I would like to offer a huge thank you to Polly who, on a cliff top in Spain, gave me the courage to risk writing this stuff down and making it public.

Introduction to the series

Hello and welcome! This is me. You and I will be sharing a journey through this book, so you may wish to know what I look like. Because practice can't happen without practitioners, I will be popping up now and again, to test-drive some of the ideas that we will be discussing.

WHY ARE THESE WORKBOOKS NEEDED?

If you are, like me, a modern-day practitioner, you are probably still dedicated to the idea of good practice, but feeling rather buffeted by many and various winds of change that are sweeping through. You are also probably feeling (like me) that it would be good to have two minutes to sit back and reflect a little: to think about what's working and what's not; and maybe even to find a little balance.

If this is how you feel, you have come to the right place. So welcome!

In this series of workbooks we will be doing exactly that, taking a little time out, thinking about what we are doing, looking at things from different perspectives and using different lenses, and trying out some practical ways of making our practice more effective, more efficient, and (above all) more satisfying.

On the other hand . . .

If you are, like me, a modern-day practitioner, you will probably also be moving far too quickly to have any time for doing anything except what you need to be doing. In other words, you probably don't feel you have time for luxuries like sitting back and thinking. Frustrating though it may be, you probably have time to do only what you *have* to do, rather than what you *want* to do.

If this is how you feel, you are still in the right place, so welcome again!

In this series of workbooks, we will be working under the clock, recognising that there are boxes to tick and targets to hit. No doubt you don't just need to keep up to date, you need to prove you are keeping up to date too, for appraisal, or for review, or for revalidation. So, as we go along, we will be providing practical examples that will help you not just to reflect upon but actually to develop your practice.

What's more, we will even be providing appraisal certificates, so our appraisers, line managers and bosses will stay happy too!

> *But you're gonna have to serve somebody, yes indeed*
> *You're gonna have to serve somebody,*
> *Well, it may be the devil or it may be the Lord*
> *But you're gonna have to serve somebody.*
>
> – Bob Dylan

WHY DID I WRITE THEM?

I have written these workbooks because there doesn't seem to be anything out there that scratches my itch. Our experience of real-life health practice is messy, complex and often chaotic. It doesn't seem to bear much resemblance to the practice we read about, or even the practice we try to teach our students and trainees.

Modern scientific and philosophical understandings of the universe are complex, messy and relational too. But our models of health and health practice often seem to be built on glib and simplistic models, or they fall into dualistic discussions (for example, about 'patient-centred' or 'practitioner-centred' care; or about 'traditional' or 'alternative' practice; or even about 'disease' and 'health'). Is the world really like that?

I have also written these books as I am worried about the levels of demoralisation and burnout among students, trainees and colleagues that I meet, right across the globe. Of course we can all get a bit tired, burnt out, and maybe even ill. If we are honest, we are often sceptical and occasionally a little cynical about what we do. But if we are even more honest than that, at heart we believe in what we do, because we think it is important.

It's not that we want to turn the clock back. We can feel a considerable (if quiet) sense of pride in how far health practice has developed. But perhaps we'd also like to think that, in the 21st century, there is a way for our practice to include and yet somehow to transcend what has gone before. It's not that we want to reject the practicalities, the science, the technology and the politics. On the contrary, I think most of us wish to accept and value them. But we also want to do what evolution always does: including, building upon and then transcending what has gone before. In so doing, maybe we can also rediscover the art of what we do, and perhaps even find a way of expressing ourselves with a little more poetry.

WHAT WILL BE IN THEM?

The answer to that is simple really. We are hoping to look at practice from different perspectives, and using different lenses, so each book takes a different view.

- Workbook 1 – *Surviving and Thriving in Health Practice*. We are the foundation of everything we do. Without us there would be no health practice. We are our own most useful tools. So, in the first book, we will look at how we can keep ourselves sharp, surviving and thriving in practice.
- Workbook 2 – *Co-creating in Health Practice*. As practitioners, whenever we come into contact with our patients, we create something very familiar but also very strange: a relationship. This relationship is neither me nor the patient, but some sort of third entity, which has an existence of its own, partly from me, and partly from the patient. This 'co-creation' is arguably our most powerful tool, but it is a tricky one to use. So we will focus on that in the second workbook, considering how we might practise in a way that co-creates healthier and happier existences, for both our patients and ourselves.
- Workbook 3 – *Turning Tyrants into Tools in Health Practice*. As practitioners we have a vast array of tools that we can use: time, computers, money, information, colleagues, equipment, targets, our workplaces and so on. If they get out of balance, however, each of these tools can become a tyrant, so that it has control of us, rather than the other way round. So in workbook 3 we will be looking at some of the most important tools (and tyrants), considering how we can stay in control of them (and not vice versa).
- Workbook 4 – *Integrating Everything*. Health practice is, ultimately, a single integrated thing. While workbooks 1–3 have been looking at the different 'bits' of this 'whole', workbook 4 is where the rubber hits the road, because it is here that we try to put it all together and come up with ways that we can integrate everything into a happier, healthier and more skilful whole within the real-life, complex and messy world of health practice.
- Workbook 5 – *Food for Thought*. We are practitioners, so we are practical, and interested in practice. So we will leave the theory until last. But most of us like a little bit of theoretical background to give context to, and to underpin our practice.[1] So workbook 5 tries to provide that. Everything that exists does so against a background. Indeed the word 'exist' means to 'stand out'. All of our experiences, beliefs and understandings of health practice derive from a living, organic and constantly moving context: whether scientific, philosophical, cultural, aesthetic, biological or spiritual. It is useful therefore to spend a little time understanding and reflecting on these building blocks of who we are. As practitioners, we don't always have time to do this, so we will leave this book until last. It will be a little luxury for those with a little more time, not essential, but hopefully a bit nourishing. Like a fireside cup of cocoa.

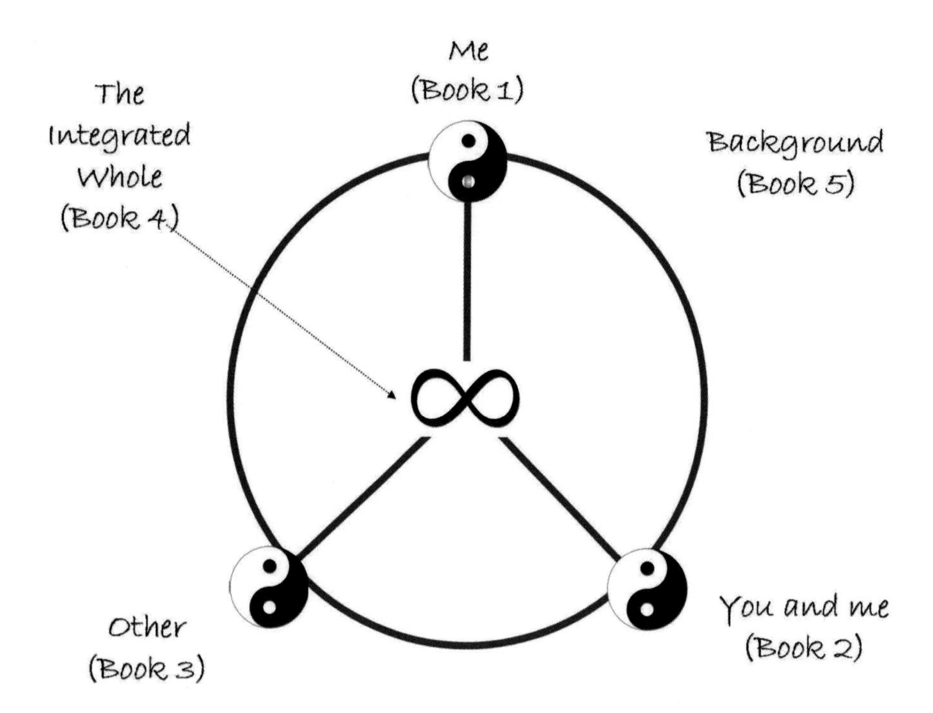

WHAT PERSPECTIVES AND APPROACHES WILL THEY USE?

In the 21st century we practise healthcare in a strange tension.

Science has taught us that we live in a highly relational, messy, multidimensional, complex, blurry and even chaotic universe. The humanities and philosophy have taught us that much of what we hold to be 'true' is relational and cultural and socially constructed. The arts teach us the value of creativity and expression in all walks of life. Spirituality teaches us about perspective, the value of awareness, and the fundamental interconnectedness of all things.

However, despite this relationality, creativity and complexity, we seem to be practising in a world that seems ever more bound and codified, with ever more targets and tick boxes, according to models that seem unrealistically geometric and two-dimensional, and with ever less room to breathe and to express ourselves.

So, in these workbooks, we will try to be practical and pragmatic. While we may not necessarily like the rules, regulations, guidelines, laws and targets that have nosed into our practice, we recognise that they have their uses. We know that health is a political football, and we are used to being kicked around a bit.

As practitioners in the 21st century we also value (and sometimes worry about) the advances that science and technology have brought. As practitioners, we are scientists, and we have a duty to do our best to ensure that what we do is as safe and effective as possible. We recognise that finding an evidence base for what we do is important not just for safety, but for development too.

So in these workbooks, we will start from the premise that we should, wherever possible, look for empirical evidence for what we are suggesting. On the other hand, we will remain vigilant to the blind spots of the empirical and technological approach, and look for alternatives to fill any gaps that we find.

Wherever possible we will look for empirical evidence for what we are suggesting.

As modern practitioners we are scientists, and also technicians, but we are artists too. There is an art to being a practitioner, and in fact practice is an art. We might lose sight of it sometimes, but we are in the business (and busy-ness) of trying to create healthier and happier existences for our patients, and hopefully for ourselves too.

So in these workbooks we will be using plenty of imagery, art and illustration to engage the more creative sides of our brains, and to remind us that integrated practitioners need to be able to find balance between creative and practical.

These days, we don't tend to talk much about spirituality. Many of us would not think of ourselves as 'religious', and some of us might be horrified at the idea that modern-day practice should have anything to do with spirituality.

But most of us perhaps like to feel that there is some purpose or meaning behind what we do. We may hope that our practice connects with and somehow reflects the values and traditions of our families as well as of our broader societies and cultures. We deal with life and death, and so with the many existential and spiritual questions that arise as a consequence. If we are to be integrated practitioners, we need to have a handle on these too.

'Along the Mystic River' – for some reason I have found myself drawn to rivers as I have written this book, so a few will be popping up as we go along.[2]

So, in these workbooks we will try to look around the edges and to peer through the gaps, asking not just: 'What should we do?' but also 'Why should we do it?' and 'What does it all mean anyway?'

Finally, we don't have to practise long to realise that there are some things that make no sense, and from which no sense can be made. Random and chaotic events, reactions and emotions may arise, surprisingly. These can be both deeply troubling but also deeply wonderful, in that they can give expression to the inexpressible. We practitioners are practical people. We like to 'do' things. But sometimes there is nothing we can do, because there is nothing to be done. At these times, we have to just 'be'. For just 'being', for making sense of nonsense, and for making nonsense of sense, there is nothing better than poetry. So we will be seeing a fair bit of that too.

Symbols and rituals are fascinating things that in some way speak to us at a 'level beyond'. It is not often easy to make sense of them, and yet we may be surprised to find that our practice is full of them.

Ars Poetica

A poem should be palpable and mute
As a globed fruit,
Dumb
As old medallions to the thumb,
Silent as the sleeve-worn stone
Of casement ledges where the moss has grow –
A poem should be wordless
As the flight of birds.
*

A poem should be motionless in time
As the moon climbs,
Leaving, as the moon releases
Twig by twig the night-entangled trees,
Leaving, as the moon behind the winter leaves.
Memory by memory the mind–
A poem should be motionless in time
As the moon climbs.
*

A poem should be equal to:
Not true.
For all the history of grief
An empty doorway and a maple leaf.
For love
The leaning grasses and two lights above the sea–
A poem should not mean
But be.

– Archibald MacLeish[3]

POINTS AND PRIZES: SOMETHING FOR NOTHING

In the initial stages of this book, my publisher explained that medical publishing is at a turning point. Whereas before practitioners might choose a book that they would enjoy reading, nowadays they are too busy for that. So the upshot is that we only read books we need to read, rather than those we want to read.

> A bit like Nanny McPhee . . .

The good news about adopting an integrated approach is we don't need to judge, we just need to adapt. If that is the way of the world, so be it, and so we have.

The particular way of the current world of health practice (at least where I currently work in the UK) appears to be a focus on objectives, outcomes, points and prizes. So the initial book has been adapted to match. Each chapter will contain activities and reflections that will meet common curriculum areas for medical and nursing practice. At the end of each book is a link to the Radcliffe Continuing Professional Development site, www.radcliffehealth.com/cpd, where you can download certificates that you can use for your CPD, appraisal or revalidation requirements.

OK, I admit it's a bit tongue in cheek, but there's no rule to say that we can't have fun while toeing the line, is there?

PROVISOS

I am, at heart, a practitioner, and a general practitioner at that. That means I am a bit of a jack of all trades, but master of none. I am partial, biased and subjective. The book is intended for all health practitioners but, inevitably, and despite my best efforts, no doubt the 'male', 'medical' and 'Western' nature of my experiences and thoughts will peep through. I hope you feel able to forgive them and look past them.

Also, I can quite honestly say that there is nothing new in this book, and I doubt there is anything in it that you could not find better argued and more coherently evidenced in other places. There is some philosophy, science, spirituality, art and poetry, but I am not a philosopher, scientist, guru, artist or poet. I am a health practitioner who dabbles.

So I have referenced those sources I can remember and can find. Others may be lost in the mists. But I do not claim any of the basic ideas in this book as my own. I have simply looked at them from my personal perspective and tried to put them together in a way that I have found useful in my own practice and in my own teaching. I hope you can enjoy them, and that you will forgive the numerous mistakes and omissions that you will undoubtedly find.

Section One

Creating in Practice

Chapter 1

The fundamental creativity of health practice

Activity 1.1: Balance, harmony, integration (10 minutes)

Reflect for a few minutes on this question:

How do I feel about the current balance, harmony and integration of my practice?

If you feel everything is OK, you probably have no need for this book.

If not, please read on. I hope it might help a bit . . .

'The Integrated Practitioner': it has a nice ring to it, doesn't it?

Who wouldn't want to feel integrated, to be integrated? In modern health practice it sometimes seems that things are going the opposite way, with increasing specialisation, compartmentalisation and regulation. With added time and work pressures we often feel pulled in every direction at once, with increasing stress, burnout and exhaustion.

'The Disintegrated Practitioner': that sounds more like it, doesn't it?

Well, yes and no. It depends on which perspective we take.

If you are a practitioner, you will be familiar with something like the record on page 15, either in paper or computer form. It's a patient record. But it is much more than just a record. Prompts appear everywhere, reminding me to check this and that, encouraging me to use drug A rather than drug B for budgetary reasons, giving me lots of boxes to tick and tasks to achieve. All useful stuff, you would think.

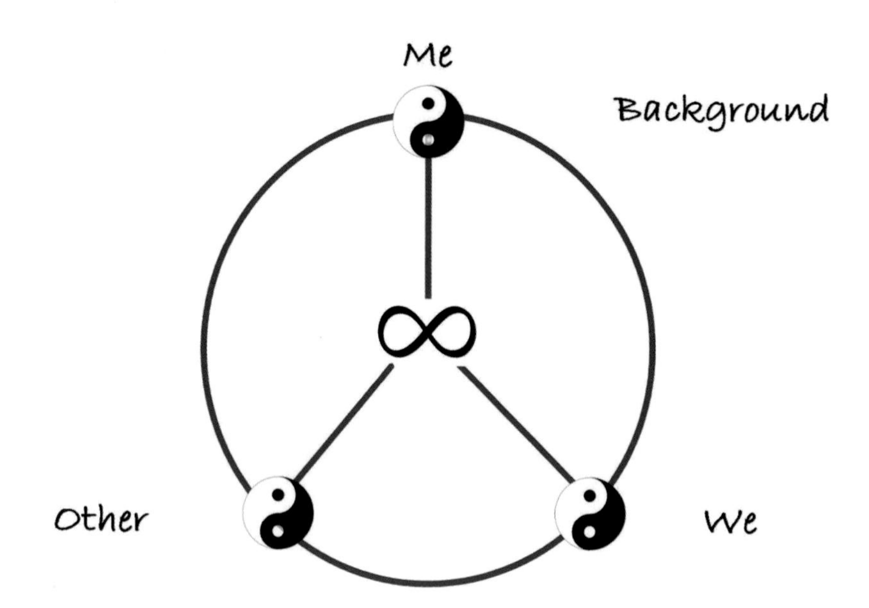

Integrating everything – our practice is the integration of all relationships: with ourselves, with our patients and with everything else.

But often it doesn't feel useful. Indeed, seeing my patients pile up on my appointment screen, knowing I will only be able to spend a few minutes with each at most; feeling under regulatory, financial and contractual pressures to hit targets and achieve goals: all this often makes me lose sight of my patient, of myself, of what we are actually trying to do together – to create better health.

> How on earth can I try to achieve any sort of integrity with all of this? Is being an integrated practitioner in any way still possible?

Well, it all depends on how you look at it.

In this series of books we have suggested that health practice can be 'seen as' many different things, and from many different perspectives. In the previous workbooks we have looked at it from the 'me' perspective (*Surviving and Thriving in Health Practice*), from the 'we' perspective (*Co-creating in Health Practice*) and from the 'other' perspective (*Turning Tyrants into Tools in Health Practice*).

In this particular workbook we recognise the fact that health practice can be many things, with many constraints and pressures. But we will also try to take a different perspective. Health practice has always been many things, with many constraints and pressures. These constraints and pressures may have changed over time and still vary from place to place, but health-practice constraints and pressures of some kind have always existed and will always exist.

Being a practitioner here and now is, from one perspective, no different to the way it has always been. It involves integration. It involves weaving together many threads into one whole tapestry. It involves taking a constrained and limited palate and painting freely. It may be a science, it may be technical, it may be psychological,

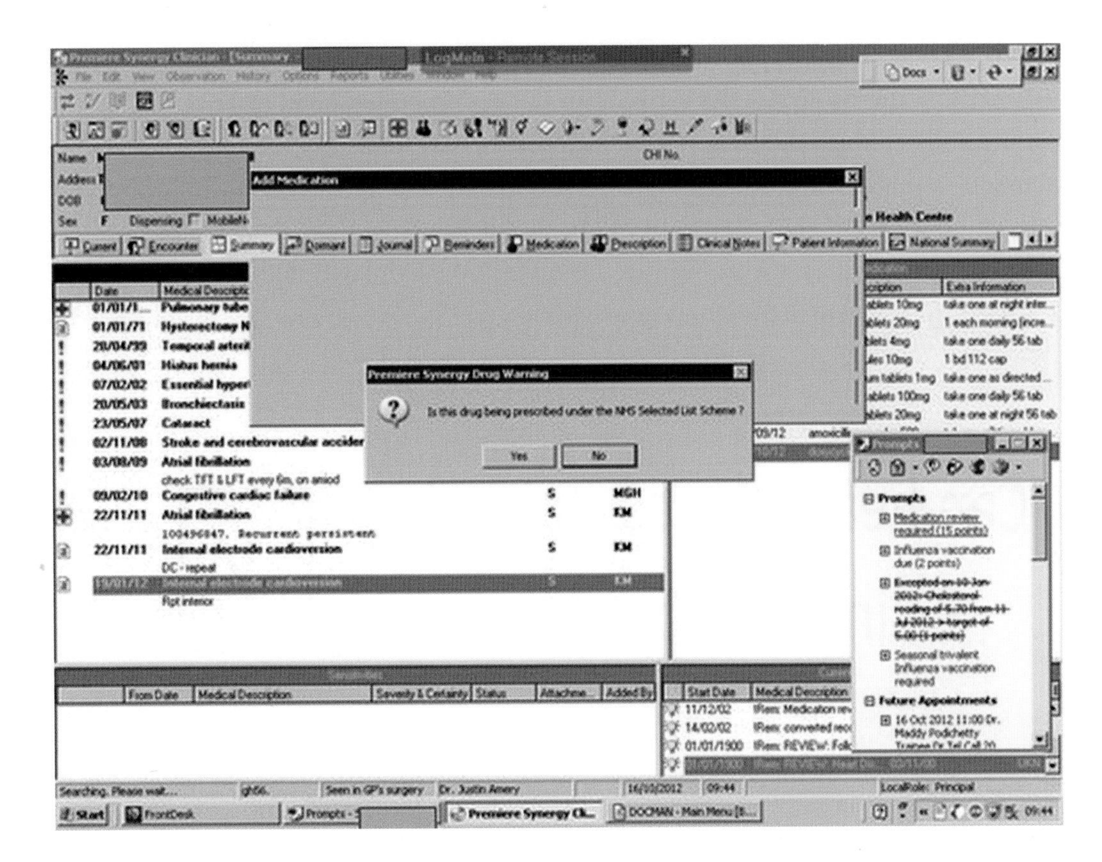

Computerised patient record from my own practice (anonymised)

it may be spiritual, but it is always an art, because it integrates everything in order to create something better.

And what we hope to create is better health.

THE ART OF PRACTICE

In practice, we create and co-create better health in an infinite number of ways, using every bit of ourselves. It is not just a technical process, though it is a technical process. It is not just a scientific process, though it is a scientific process.

It is always a complete expression in which we can choose to touch, see, hear, think, feel, sing, narrate, transfer, model, ritualise, dance, and even play.

As we dance (or sing, or play . . .) together, we pull all these various threads of information together within our imagination, and using that imagination we co-create an actual, new and healthier existence for our patients, and hopefully for ourselves too.

If we are practising in an integrated way, we can heal ourselves as we heal others.

All works of art start in the imagination. We imagine what we want to create, and then we go about trying to create it.

It's the same in health practice. Without the imagination to put ourselves in the shoes of our patients, to see the world from their perspectives, and to imagine how their life and health could be 'better', we will not be able to use all the tools we have discussed to this point constructively, we will not be able to integrate them effectively, and the 'health' that we create will be less perfect.

Without imagination we are simply computers. We can analyse the data but we can't use it to create and construct new realities.

Can health be contained in a tick box?

A QUICK RECAP

If you've not read workbooks 1, 2 and 3 (and maybe even if you have), all of this might come as a bit of a surprise. If so, here is a quick recap.

If it's not new to you, please skip on to the next section.

The universe is fundamentally relational, complex and infinite. As part of that universe, we should not be too surprised to find that we are fundamentally relational, complex and infinite too. The fact that we are self-conscious beings makes our relationality, complexity and infinity even more relational, complex and infinite.

Actually, strictly speaking, you can't get more infinite than infinite.

Fine, but you get my point. The key thing is that despite all this, and despite the fact that we exist on two levels (both as physical beings within the physical universe and also as conscious beings within our conscious minds) we also exist as one integrated whole.

So, how can we exist at two levels at once, as complex, relational, infinite entities and also as simple, single, integrated wholes?

The answer is that, unlike most other things in the universe, we are conscious. Our consciousness has an extremely rare, if not unique, quality. It is self-creating. We are never not conscious of being conscious. We can only ever be conscious of being conscious (of being conscious, of being conscious etc. etc.).

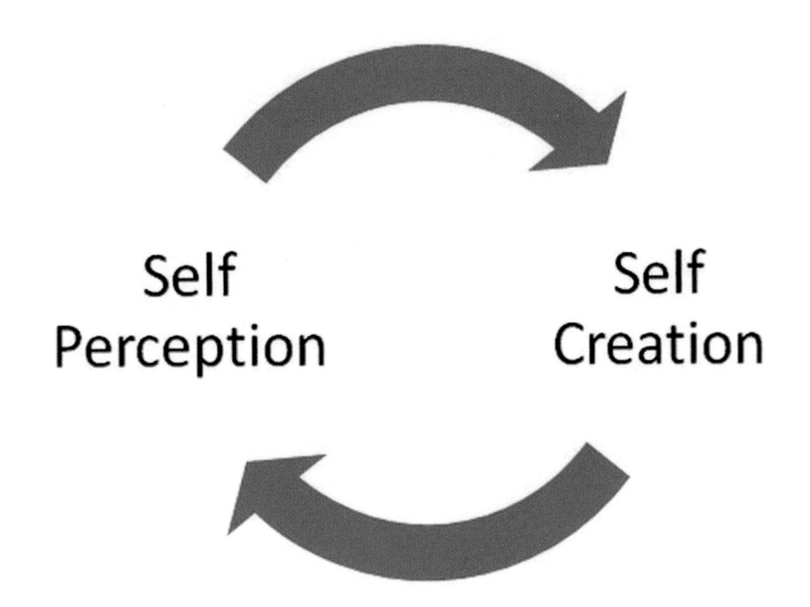

Self-consciousness is a product of our self-perception, which is experienced through our self-consciousness

Our experience of life, of our existence, is therefore something we create within our own consciousness. Consider the riddle of the red rose: are the qualities of being red, and a rose, qualities that exist within the rose, or within our consciousness, or somehow both? What happens when the lights go out?

Question: Is a rose red at night?

What we create depends on what we perceive. In a relational universe, things look different depending on your perspective. What one person sees as 'health' and 'practice' may be quite different to another. So our job as health practitioners will vary depending on the perspectives we take, and the perspectives our patients take.

Rather than try to define the undefinable, or to pin down the ineffable, we have suggested that a useful perspective might be to see 'health' as a form of integration and harmonic balance between all the relationalities of 'existence'. It then follows that 'health practice' may be seen as the attempt to help patients to achieve this integrated, harmonic balance.

As we have seen in the previous workbooks, we can only experience the relationalities of existence from one of three perspectives – as an individual experience of 'me', as a shared and co-created experience of 'we', and as an external and remote

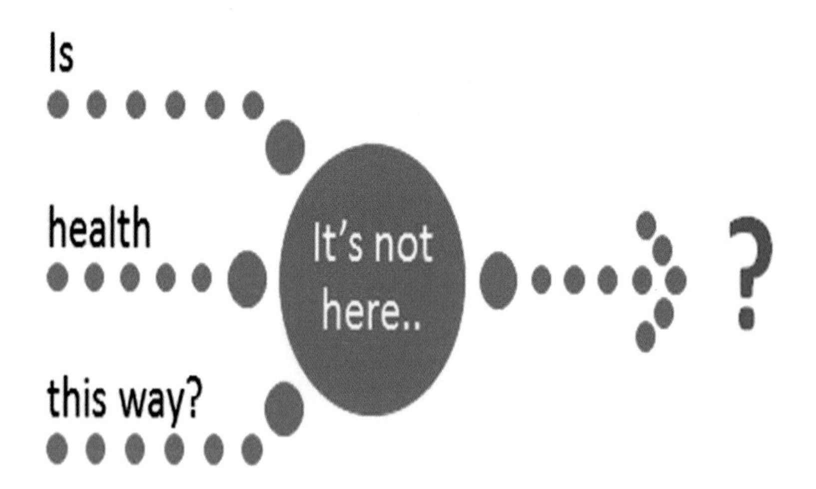

experience of the 'other'. These three perspectives give us three different sets of tools that we can use for our creation.

1. The most important tool that each of us has is ourselves. Without us there would be no health practitioners and so no health practice.
2. But there are many, many 'other' tools. These others can become tyrannical if we attach too much importance to them but, if we can keep perspective, we can use them skilfully to add texture, depth and quality to our creation.
3. Finally, because we can only create health in partnership with our patients, they are our tools too (just as we are their tools). We have to work skilfully together with our patients not just to create, but to co-create, better health.

But we don't experience these things neatly and separately. We experience them as one integrated whole, in the moment, and moment by moment. Working with the palette of what we perceive, our consciousness creates our existence, just like an artist.

That means we now need to pull together the perspectives of the first three workbooks. It means we need to look at how we integrate everything into one, infinite whole; and so create one, integrated, harmonically balanced thing: health.

And that is what we shall try to do in this workbook.

Crabbit Old Woman

What do you see, what do you see?
Are you thinking, when you look at me –
A crabbit old woman, not very wise,
Uncertain of habit, with far-away eyes,
Who dribbles her food and makes no reply
When you say in a loud voice,
I do wish you'd try.
Who seems not to notice the things that you do
And forever is losing a stocking or shoe.
Who, unresisting or not; lets you do as you will
With bathing and feeding the long day is fill.
Is that what you're thinking,
Is that what you see?
Then open your eyes,
Nurse, you're looking at me.

– Phyllis McCormack[4]

A nurse's response to 'Crabbit Old Woman'

What do we see, you ask, what do we see?
Yes, we are thinking when looking at thee!
We may seem to be hard when we hurry and fuss,
But there's many of you, and too few of us.
We would like far more time to sit by you and talk,
To bath you and feed you and help you to walk.
To hear of your lives and the things you have done;
Your childhood, your husband, your daughter, your son.
But time is against us, there's too much to do –
Patients too many, and nurses too few.
We grieve when we see you so sad and alone,
With nobody near you, no friends of your own.
We feel all your pain, and know of your fear
That nobody cares now your end is so near.
But nurses are people with feelings as well,
And when we're together you'll often hear tell
Of the dearest old Gran in the very end bed,
And the lovely old Dad, and the things that he said,
We speak with compassion and love, and feel sad
When we think of your lives and the joy that you've had,
When the time has arrived for you to depart,
You leave us behind with an ache in our heart.
When you sleep the long sleep, no more worry or care,
There are other old people, and we must be there.
So please understand if we hurry and fuss –
There are many of you,
And so few of us.

– Anon

Activity 1.2: Draw health (10 minutes)

Take a blank piece of paper and draw health, creatively.

Keep your sketch. We will be coming back to it later.

Chapter 2

Health as a creation

Activity 2.1: What am I creating? (20 minutes)

Think of your own attitude. How creative do you feel? Is creativity something that excites or scares you, or maybe a bit of both?

Think of yourself. Do you think of yourself as creative? If not, why not?

Finally, think of your practice. What are you creating with your patients? Is that creation real – for your patient, for yourself, for others?

The core suggestion of these workbooks is that health can be seen as the creation of integrated, harmonic balance. It follows then that health practice can be seen as a process of creation of integrated, harmonic balance. In other words, health practice can be seen as a fundamentally creative process.

In workbooks 1, 2 and 3 we have looked at the various tools, relationships and perspectives we can use in our creation. In this book we are going to try and pull these all together and attempt to start creating something truly worthwhile: better health.

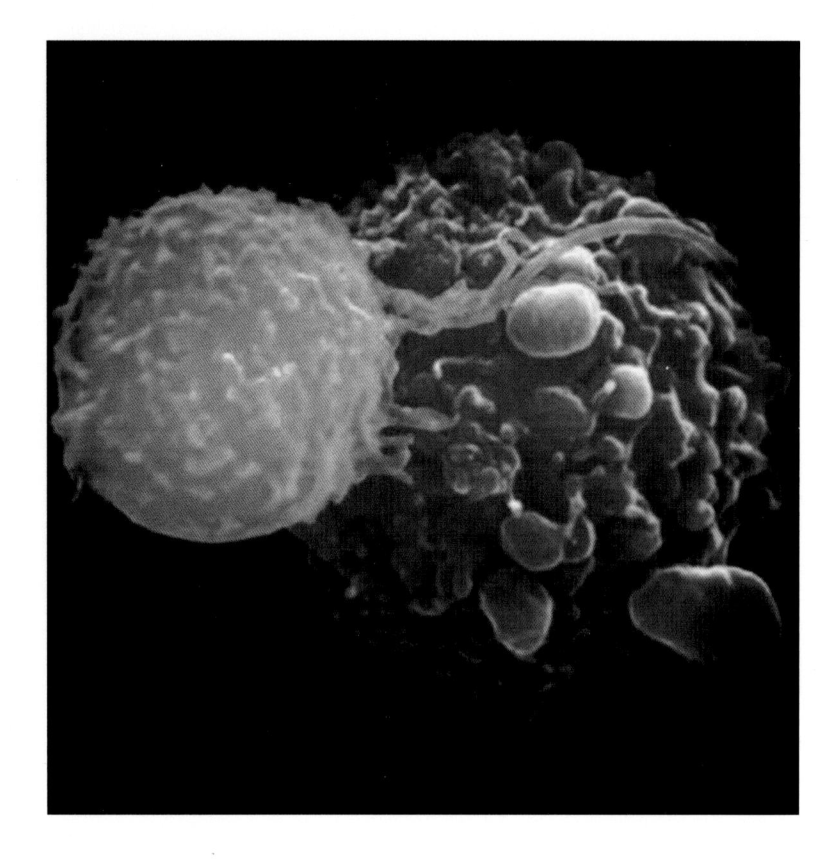

Creating better health[5] – a healthy immune system is a fully integrated and balanced creation. Without it we would be overcome by infections and cancers. How can we improve that integration and harmony?

THE CREATION: BETTER HEALTH

Creation is the practice of bringing something new into being.[6]

As we have seen in these workbooks, health is not something that has a concrete and separate existence. We can't see it as it walks by. It is something we experience within our consciousness, and so the word 'health' has little meaning outside consciousness. It is a product of consciousness.

Health is therefore not something that is already there, something that we can go out and find. It is something we create in our consciousness, and co-create in each other's consciousness.

That is not to say health practice is not scientific. On the contrary, we can use empirical, scientific approaches to discover a great deal about health, the causes of health, and the tools we can use to improve health.

It is also not to say that health practice is not technical. On the contrary, health practice requires great skill and technique: physical, psychological, social and maybe even spiritual. Without these technical skills, and without the technology that supports them, we would be much less effective.

But it also means that health practice is fundamentally creative, because we create better, healthier states of existence, both for our patients and ourselves. That means we are also artists too.

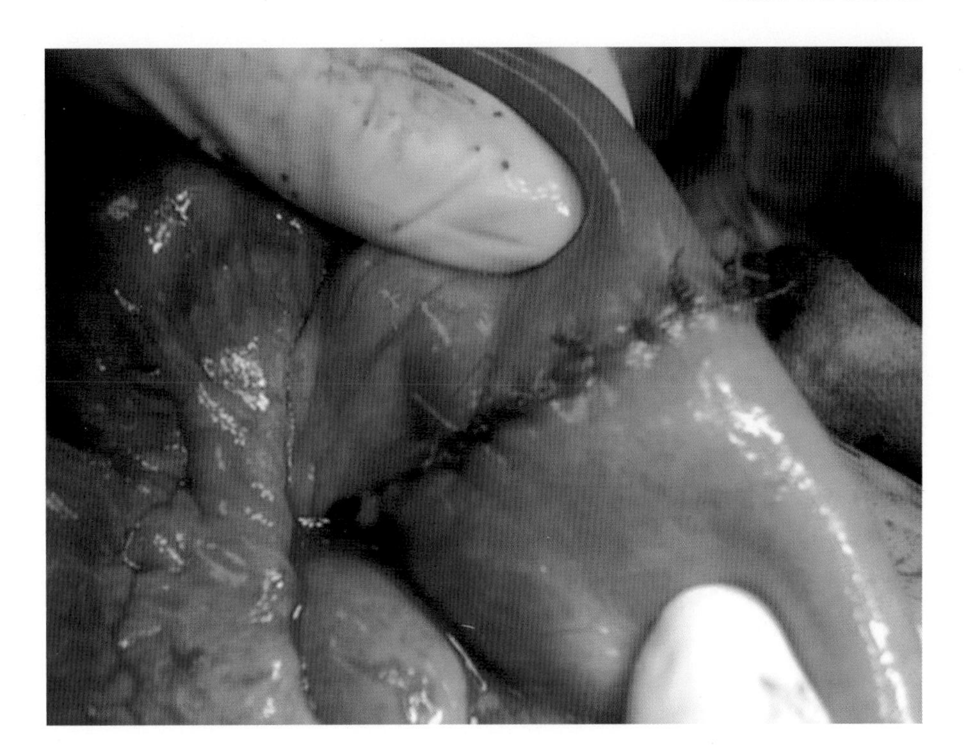

Creating better health[7] – only a very few people in the history of humankind have practised enough to become expert enough to rebalance and reintegrate the anatomy of the human body once it has become damaged.

Art is work.

– Milton Glaser[8]

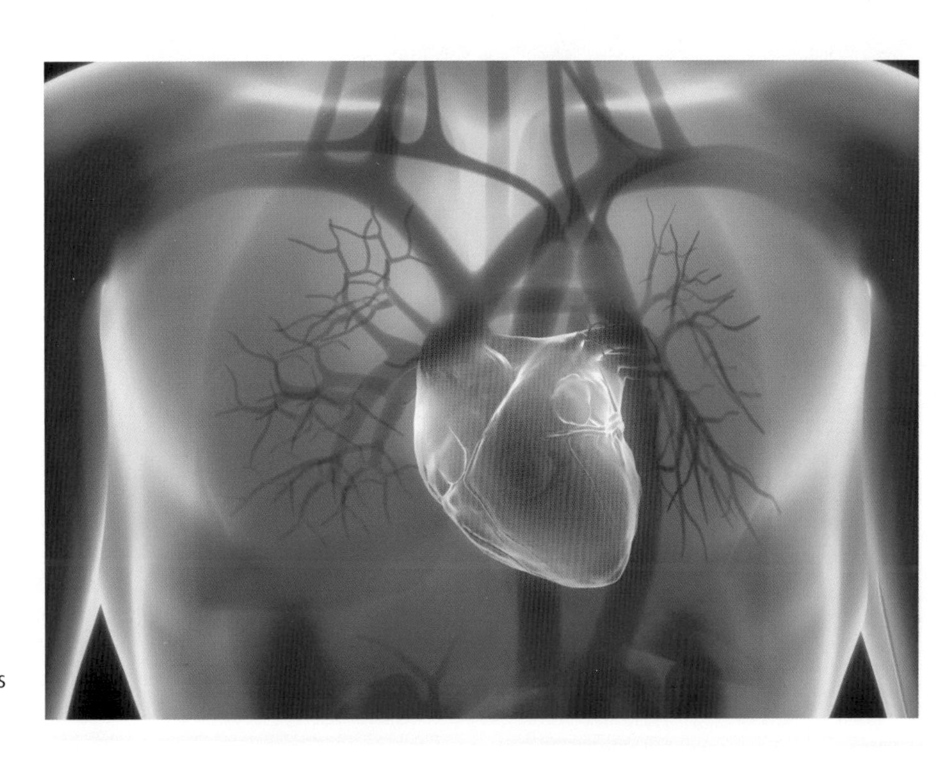

Creating better health – the human circulatory system[9] is both amazing and beautiful. To maintain and repair it takes tremendous skill.

PRACTITIONERS AS ARTISTS

In the current world of science, technology, evidence, regulations, guidelines and targets, the idea of health practice as creative art may seem unlikely, incredible even. But let's take another perspective for a second.

We may think of art as being entirely inspirational, and we may think of artists as being creative and other-worldly. And that would be right. To be creative we need to be inspired, to have an eye on the world of possibility as well as the world of actuality.

But creativity isn't just about that. To be creative, we have to put our ideas into practice. And putting ideas into practice takes fairly hard-nosed skills: clarity of thought, logic, strategy, energy and dogged determination.

As practitioners we have all of those in our locker, don't we?

They say brilliant art is 99% perspiration, 1% inspiration. Are we any different? We have certainly put in the perspiration, so is there any reason we should not have the inspiration too?

Creating better health – this little boy lost his eye to cancer, and in the process lost his social life to shame and stigma. A volunteer helped rebalance things, and helped him reintegrate himself, by giving him these sunglasses.

OUR PALETTE: ALMOST ANYTHING

In an infinite universe, being a practitioner can mean being many things and there are many ways to practice.

We might be trained in any one of many different approaches or techniques. We may be prescribers, or surgeons, or nurses, or counsellors, or physicians, or carers, or hypnotists, or masseurs, or acupuncturists. The list goes on. No one can do everything. And in today's complex and specialised world, even if we tried to do everything, we'd probably end up doing nothing well. Whatever we do, we are constrained in some way or another.

But constraint is not necessarily a barrier to creativity. There are an infinite number of points between 1 and 1 000 000; but there are also an infinite number of points between 1 and 2.

And that's a lot of points to diverge from and converge to . . .

So we should not be worried about having one particular speciality, or one particular expertise, or working in one particular location. We still have plenty of room to create. What's more, the more expert we become in an area, the more creative we become in that area too.

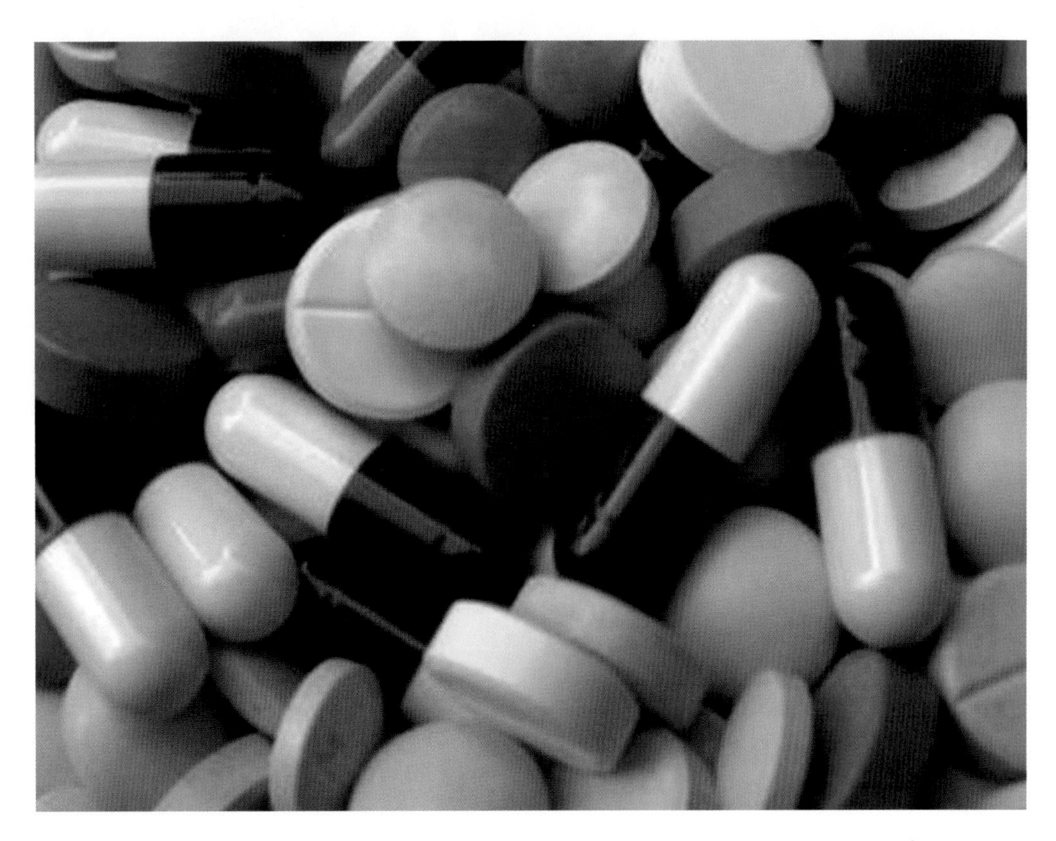

Creating better health – sometimes our molecular balance becomes disrupted, and we can use drugs to reintegrate and rebalance ourselves. We may not think of drugs or other treatments as tools for creativity, but why not?

Creating the conditions to enable creativity to happen is harder than it sounds, especially in the hurly-burly of modern health practice. But that is no excuse. Our patients need us to be creative, because they want us to co-create better health with them.

So in the rest of this workbook we will have a think about how best we can create the best conditions in which to create. You can see there is a certain self-reference involved in this, but that is not surprising, because we are self-referential, and self-creating, beings.

Creating better health – health is experienced within, and as part of, our families and social groups. Helping families to become better balanced and more integrated can create better health for the individuals and for the whole.

Blue sky, but a hint of autumn chill
Muffled roar of distant traffic
A yellow leaf spirals silently down
From the willow under which I am standing
Later I see movement in a line of trees
With binoculars I see the brilliant blue
Flash of the jay's wing feathers
Exotic and almost tropical

– Tom Nicholson-Lailey (Tom is a GP, colleague and friend in Oxford)

Activity 2.2: Creativity (15 minutes)

Revisit Activity 2.1.

Consider how you would like to be more creative within your practice and, if so, how you could do so.

Chapter 3

What is creativity?

Activity 3.1: Revisiting creativity (10 minutes)

Consider the picture of the Taj Mahal below. Have a think about the processes involved in the creation of such an amazing, and enormous, piece of art.

Being creative isn't the same as being airy-fairy or arty-farty. A baby can create, but what he or she creates is usually a mess.

We are skilled practitioners, and so what we create can be highly skilful. But to create a skilful creation, we need to balance opposing forces – the drive to open up and the drive to close down, the need to start and the need to end, the ability to imagine in our minds and to enact that idea with our bodies.

Each of us is naturally creative – we are creating ourselves right here and now – but each of us tends to go in one direction or another. Finding balance is tricky.

Creativity is what happens at the nexus between our consciousness and our physicality. As we have seen in these workbooks, because we are conscious we exist simultaneously at different levels: the 'physical' levels of energy, forces and matter; and the 'conscious' levels of colour, texture, ideas and emotions. To be alive is to integrate and to balance all of these different levels into one integrated, self-created whole.

In response to data from the physical level, we create internal representations of what we perceive, and then imagine new ways of thinking or acting in response. This interplay between our imagination and action is the basis of creativity: we imagine something new, then we try to make that new thing happen. It comes to us so naturally that we don't necessarily think of it as creativity, but it is.

Health practice follows the same creative process. When patients come to see us, we communicate with them physically, verbally, non-verbally and in many other ways. Through this communication we obtain a wealth of sense data. We then integrate this data into our existing internal representations; searching always for what 'fits' and what doesn't 'fit' with what we 'know'. From here, we can begin to imagine what it is like to be the patient, what we could do improve his or her health, and how we might go about making that improvement happen. We then act to try to put our ideas into action, thereby co-creating 'better health' with our patients.

As we can see, there is a variety of skills required in this creative process. These include the technical skills of communication and design and the scientific skills of assessment and analysis. But they also include the artistic skills of imagination and creation.

To be creative we need to be able both to diverge and to converge. We have to be able to diverge from the moment, exploring possibilities and generating ideas. We also have to converge on the moment, planning and modelling our creation and then making it happen in practice.

So creativity is not an optional extra to health practice. It is an integral part of it.

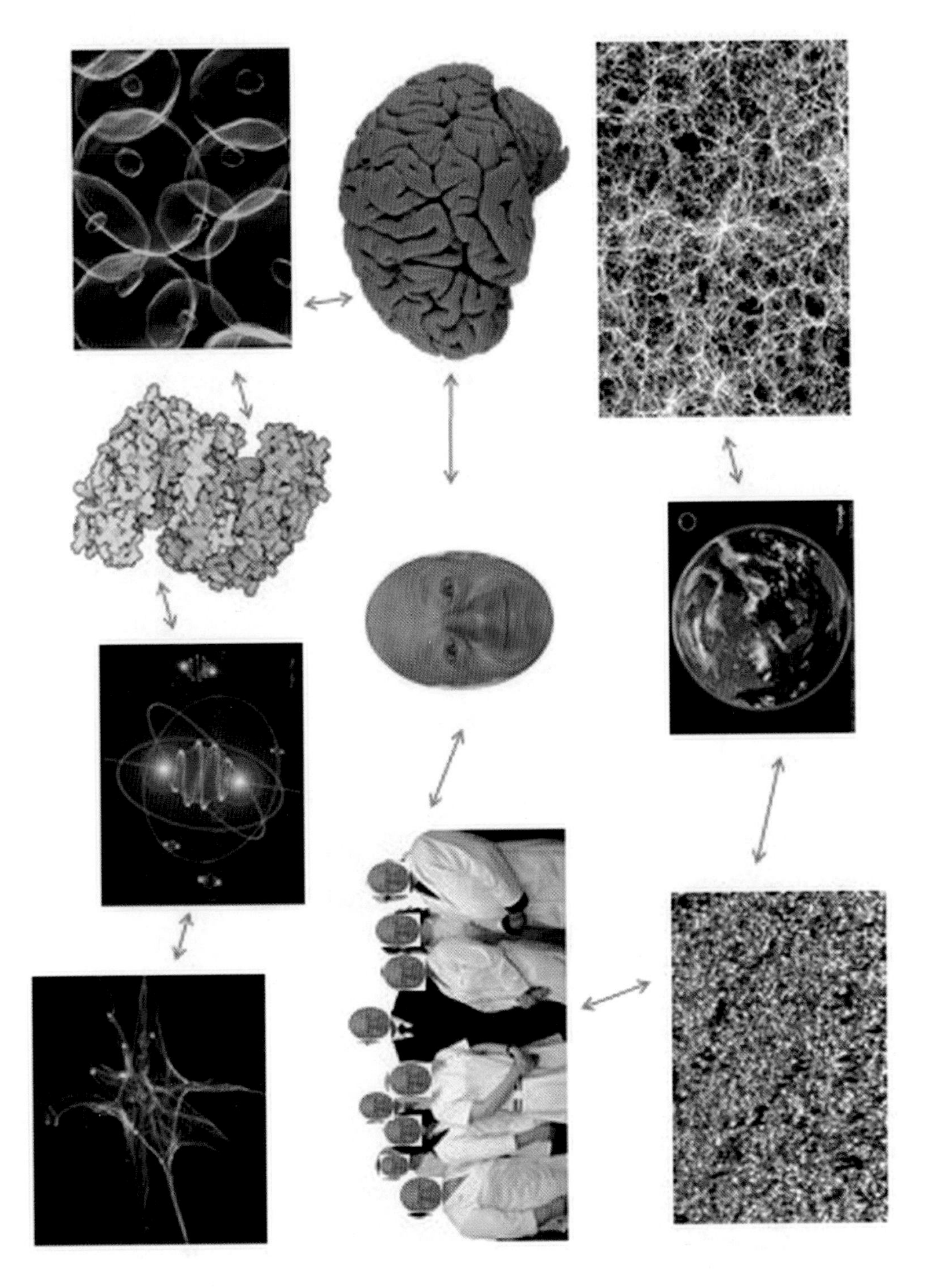

As conscious beings we exist at many levels at one and the same time; for example, at sub-atomic, atomic, molecular, cellular, organic, individual, societal, planetary and universal levels. Each one of these levels simultaneously includes and transcends each previous level, exists at its own level, and forms a part of higher levels. These levels are called 'holons'. Health practice involves integrating and balancing all of the holons creatively.

IMAGINATION

We are all familiar with the sense of an idea coming to us out of the blue, like a light bulb switching on, the 'eureka' moment. This can lead us to assume that creativity is a rather arbitrary, even magical thing, which just 'comes upon us'.

As practitioners, we want to be practical, and magic is not practical. If we are led to believe that creativity is mysterious and unknowable, it ceases to be of any practical use: if we can't learn it, we can't use it.

But in fact the science and psychology of creativity does not suggest that creativity is a gift from the gods. It appears to be something we can learn, practise and master, just as we can learn, practise and master the science and the technique of health practice.[10] What is more, as we shall look at later in this workbook, the evidence suggests that the more expert we become in health practice, the more creative we become too.

Imagination is a process of *divergence*. From the specifics of the moment, we 'zoom out' in many different ways, exploring the world of possibility, sifting and testing each idea against our experiences and our memories, gradually honing them down until a particular idea suggests itself.

This is what is called divergent thinking. We may not be aware we do it, because it is largely subconscious. The process of generating ideas, sifting and testing happens

Eureka – divergent thinking

out of sight, and out of mind. It only becomes conscious when the idea 'pops' into our conscious mind, as if from nowhere. But this is just an illusion. Imagination is simply another form of thinking.

Divergent thinking is relaxed, effortless and largely subconscious. To encourage divergent thinking, the science shows that we need to quieten down the rest of our brains, to relax, and to give ourselves time and space. This space allows our subconscious to quietly imagine and create, and enables our consciousness to recognise that still, small voice of creativity when the 'good idea' seemingly pops up. That's why we have put so much emphasis on relaxation, mindfulness, clearing and awakening in these workbooks.

ACTION

Being creative is not just about dreaming up good ideas. To be fully creative, we have to put our 'good idea' into action and create the thing that we have imagined. This is a *convergent* process, in which we zoom in, choosing one idea from the many, analysing it for logic and validity, mentally modelling and testing it for practicality and applicability, and then putting the idea into action.

Convergent thinking tends to be conscious, logical and effortful. It builds logically from a start point to converge on an endpoint. We are aware we are doing it and it feels like hard work, because it requires concentration rather than relaxation.

Thinker – convergent thinking

Convergent thinking is as essential as divergent thinking for creativity because it is systematic, logical, pragmatic and strategic enough to choose the right action at the right time, and to complete that action skilfully, efficiently and effectively. To encourage convergent thinking, we need to be alert, focused, armed with the right information, and to have protected, uninterrupted time and space to think carefully.

CAN WE TRAIN TO BE MORE CREATIVE?

The science and philosophy of creativity (which we will cover in more detail in workbook 5) seems to suggest we have to try to achieve a balance between divergence and convergence, between opening up and closing down, between imagining an idea and then making that idea happen. Divergent thinking tends to be subconscious, effortless, and intuitive. But it requires a peaceful mind and a calm body. Convergent thinking is conscious, logical, sequential and effortful. It requires time, concentration and hard work. To be health practitioners we need to do both, often at the same time, often switching rapidly from one form of thinking to another.

The good news is that most of our training already encourages convergent thinking. To be effective health practitioners in the current world, we have to think quickly, efficiently and effectively, even with time pressures and interruptions. Much of the development of health practice in the last century has focused around convergence: enabling us to process ever larger amounts of information, to become ever more specialised and expert in our analysis and modelling, and using technology to enable ever more sophisticated action.

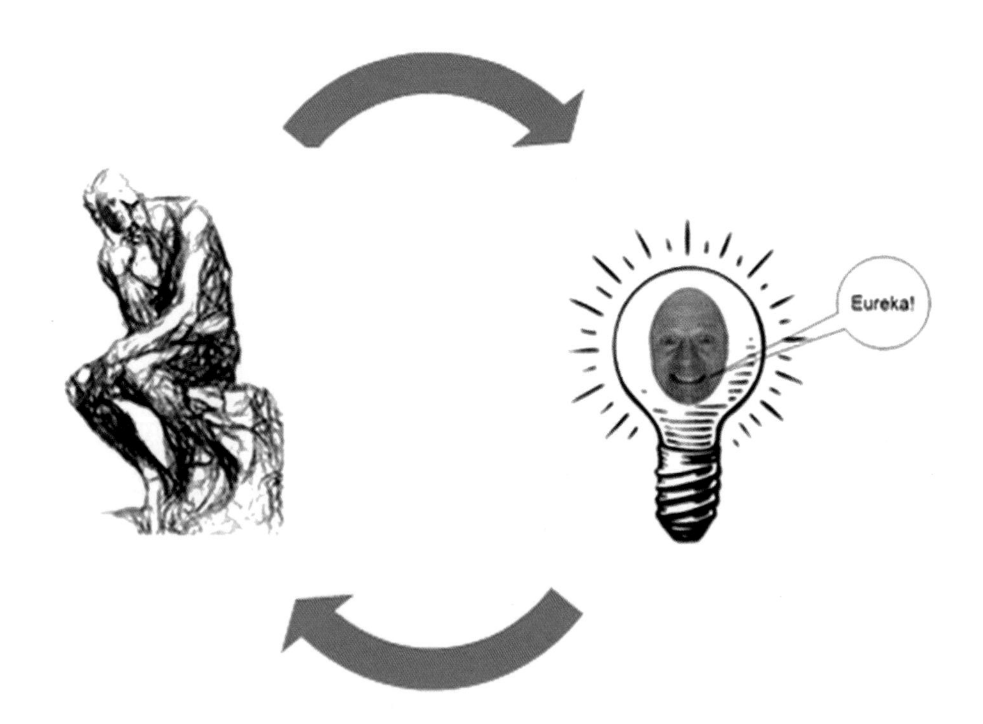

Converging and diverging – balancing and integrating divergent and convergent thinking.

The other good news is that divergent thinking comes naturally. We are naturally creative, because we are conscious, and consciousness is a fundamentally creative process. We have all had the experience of good ideas popping into our heads, maybe when we are having a relaxing shower, or while taking a walk, or when waking up in the morning. So we don't need to train ourselves to diverge, we just need to create the conditions to allow it to happen, and to trust ourselves when it does.

FACTORS THAT MAY AFFECT THE WAY WE THINK

It is common to hear people say, 'I am not creative.' Hopefully, we can now see that we all are, because we are all conscious, and consciousness is a fundamentally creative activity and state of existence.

So what lies behind this common misconception?

Perhaps it is that we tend to confuse our creativity with our preferred style of thinking. We each tend to have a natural 'preference' for either convergent or divergent thinking. Sometimes we confuse 'divergence' and 'imagination' with creativity. But it is not that people who prefer divergent thinking are more creative. To create we need also to put ideas into practice – to be convergent too.

So while it is wise to be aware of one's own preference for divergence or convergence, to be truly creative we need both. Each of us needs to be able to cultivate our least preferred style, so that we are each able both to diverge and also to converge competently, and so create effectively.

Fortunately it appears that there are tools we can use to help us foster both divergence and convergence.[11] Overall, the main influence on the way we think seems to be the degree of 'cognitive ease' we feel.

In general, factors that increase our sense of 'cognitive ease'[12] will make us more likely to think divergently, whereas factors that increase our sense of 'cognitive threat' make us more likely to think convergently, but there are other factors too.

- The environment: busy, noisy, red-coloured and stressful environments will prime us to converge whereas peaceful, quiet, blue-coloured and relaxing environments will prime us to diverge.
- Familiarity and expertise: the more we feel expert and familiar with the situation, the more easily we will be able to diverge.
- Substances: stimulants such as coffee, chocolate and amphetamines will make us more effective at converging but less effective at diverging. Relaxants such as alcohol and cannabis will do the opposite.
- Question framing: if questions and issues are framed in a closed way (e.g. where we are given specific algorithms, guidelines or targets) we will be primed to think more convergently than if they are framed openly.
- Mood: if we are feeling high, we will find it easier to diverge than if we are feeling low, when we will be better at converging.

There does not seem to be any evidence for this, but it would not be a great leap of faith to assume that, in modern health practice, there are far more primers of convergent thinking than of divergent thinking, for example time pressure, burnout, exhaustion, resource pressure, targets, guidelines, busy and threatening environments, information overload, rules and regulations.

> This also fits with my experience of teaching and training consultation skills; in which practitioners more commonly run into problems being too convergent (particularly 'paralysis by analysis' and becoming 'closed down' to the patient) than in being too intuitive.

We sometimes forget how fast and effective intuition can be. Given that we use it in scenarios where we are experienced and expert, it is pretty safe and effective too. As intuition is both much faster and much easier than reasoning, it seems a shame if we are nervous about using it.

THE FLOW STATE

Probably the best state of existence for creativity is the 'flow state'. We covered this in workbook 1, but, as a brief reminder, flow states happen at the junction of arousal and control, of anxiety and relaxation, of worry and boredom, and (of course) of challenge and opportunity.

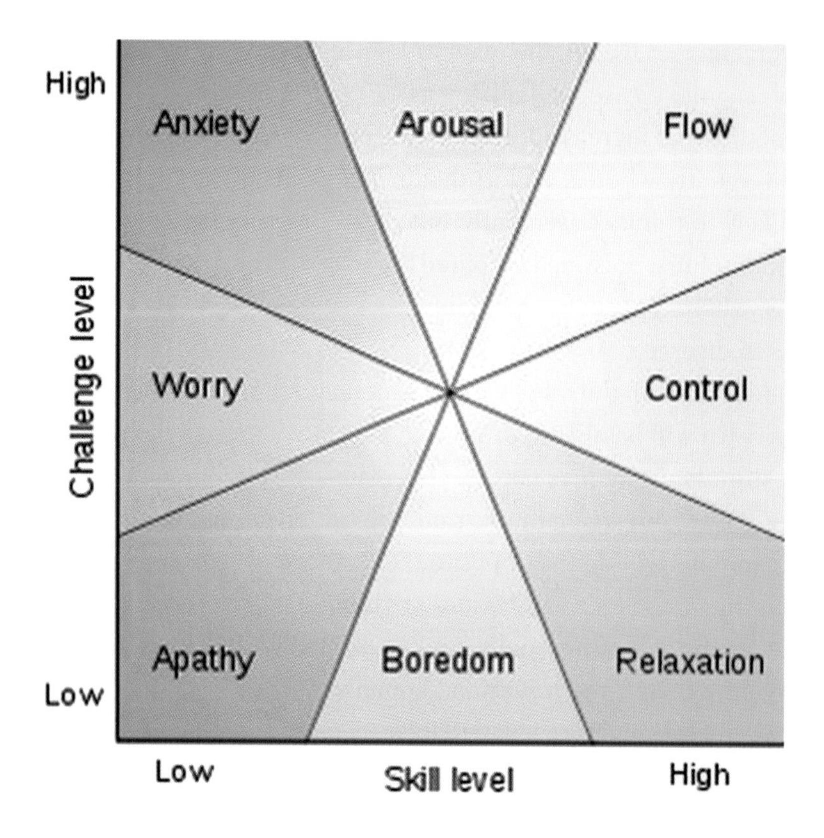

The flow state as a balance

Jimi Hendrix in a flow state[13] – a perfect balance of technical skill and creative expression (although the LSD may have had an influence too)

The flow state is probably the most useful state to be in for creativity (and therefore for creative practice), as all our focus and effort is single-pointedly targeted on the task in hand. It is therefore an ideal state for us to try to achieve in practice.

INTEGRATING CREATIVITY INTO OUR PRACTICE

Integrating creativity in practice is fundamentally about trusting ourselves and letting go.

What we create, better health, is a brilliant creation which takes great expertise and devotion. To create it, we have to become aware of, balance and integrate many diverse threads. But if we can become mindful of ourselves as creators, mindful of the tools at our disposal, and mindful of the importance of keeping these tools sharp, there is nothing to stop us creating expertly.

In order to improvise, to head off the page, we need to know what's on the page first. No matter how brilliant we may be, we won't be as creative without hard work and plenty of practice. While anyone can be creative, we are more likely to be effective in our creativity in areas where we are expert. So becoming more expert or specialised does not act against creativity, it promotes it.

In order to imagine, we need to create space and to get out of our own way. If we allow ourselves to become blunted or burnt out, or if we become tyrannised by all the various pressures of practice, or if we fail to communicate effectively with ourselves and with our patients, we will be constrained, and create less effectively.

So, at the actual moment of creation, we aim to let go and to trust ourselves to create the right thing for that particular moment. We can trust ourselves to let go and to create, for all the reasons that we have covered in these workbooks.

- We know what we want to create: better health.
- We have our palette: everything that we have learnt, from these workbooks, from our practice, from our lives. All the 'me', 'we' and 'other' factors of health practice, of existence.
- We have ourselves, the artists: prepared, dedicated, mind and body cleared, thoughts converging and diverging, right and left brains switched on, standing and ready.
- We have our canvas: this moment, in practice, right here, right now.

But before we come to that, we need to back up a bit, and look at nothing.

The Fountain

Swimming deep in the rippled sky
Water holding me in
And within
I
burst high
Cascading splintered light
Fractal faces of nothing
as the rainbow begins
and ends
In me

– JA

Chapter 4

The infinite world of practice (and the 'no-model model')

Activity 4.1: The infinite world of practice (30 minutes)

Think back over your entire life as a practitioner. Have you ever experienced two identical moments?

Think forward to the rest of your life as a practitioner. Will you ever experience two identical moments?

When you next see a patient, keep a little bit of your mind back. As you follow a particular path with that particular patient, allow that little bit of your mind to play around, and to imagine all of the alternative paths you could have taken, or could take.

Consider the differences between models and real-life practice. Reflect on the world of possibility and plurality that you practise in.

Nothing happens next.

– Zen saying

THE IMPORTANCE OF NOTHING

Let's start with a very hard concept. Don't worry if it doesn't make sense just yet. Sense will probably emerge as you read through this book. Even if it doesn't, we will use other perspectives and routes that will arrive at the same place.

The hard concept is nothing.

It doesn't exist.

Everything that exists, that has been created, exists out of nothing, by definition.[14] Without nothing, there can be no-thing. This may seem like an obscure philosophical point. But it is absolutely crucial to creativity, and so to health practice.

Let's just rewind a little.

As practitioners, we want to practise. As professional practitioners, our lives are full, our minds are full and our hands are full. We have patients to care for, colleagues to meet, bosses to please, exams to pass and boxes to tick. As it is our families hardly see us, and when they do we're tired out. Sometimes there really does not seem any time, or any point, in trying to be more.

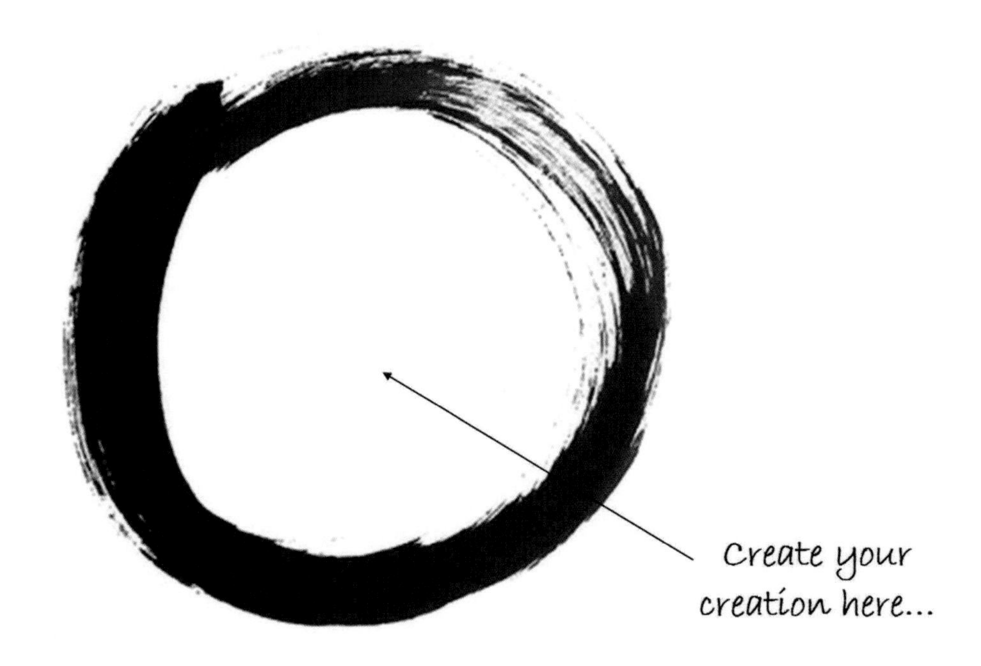

The 'enso' symbol symbolises the emptiness and thus the potentiality of the moment, because we need space for nothing if we are going to create something. Our practice may be full of 'some-things', and those 'some-things' may entirely capture our attention and our action. But if we want our practice to be creative, it is important to become aware of the emptiness and potentiality that underlie everything, our practice included.

But let's just pause for a second. Is being full really such a good way to be, in order to practise? If we have no time, no energy and no space, how will we create better health? And if we can't create better health, what are we here for?

It's not that the lessons we have learned are not important, or the training invalid, or the experience useless. There are five workbooks because there is a lot to learn.

And they only scratch the surface. So, on the contrary, knowledge, experience and expertise are vital to effective practice. Each lesson we learn helps us get more effective and more efficient. But, at the moment of practice, we cannot focus on these things.

It's not that we should become other-worldly and renounce the knowledge, or the science, or the techniques, or the technology. These are crucial and useful tools for practice, but they are not the practice itself, they are not the creation.

It's not that we should pretend that we are not busy, nor exhausted, nor that we are free from constraints. We are busy, exhausted and constrained a great deal of the time. We practise in real life, in the world of the actual. We can't change the world. We can't change the pressures, the demands, or the targets.

But we can change our perspective, and changing perspective changes everything.

Let it go. Let it all go. Trust yourself.

We are scientists, we are technicians, we are healers, but we are also artists, because we create. We create better health.

We have everything we need. We may want more, but we have enough to create better health, right here and now. The thing which is often missed, which we often forget, is the importance of creating emptiness, creating nothing, before we can fill the emptiness with our creation of something. In other words we have to find nothing before we can create something, because everything comes into being out of nothing.

By definition.

Look again at the Enso, and the emptiness. It's not telling us anything. It's inviting us.

PRACTICE AS AN INFINITE ENTITY

Here's another hard but important concept: infinity.

As well as being fundamentally empty, our practice is also fundamentally infinite.

We are physical as well as conscious beings, so our practice is bounded by the laws of the physical universe. But, even within those boundaries, the ability of our consciousness to create, and to co-create, is unbounded, limitless. Within the boundaries of space and time there is still an infinite number of possibilities.

So when we step into the next moment of our practice, we step into a moment of pure potential, capable of an infinite number of arisings, flavours, textures, happenings and outcomes. And that is why, however long we practice, we will never experience one moment that is exactly like any other.

We are practitioners so we normally hope to create, and co-create, something that is 'healthy' – healthy for our patients and healthy for us. In these workbooks we have seen that there are many, many ingredients and techniques we can use to create: our self-creation, the patient's self-creation, the co-creation we create with our patients and with all the myriad 'other' entities and relationships that we have.

Altogether that is a tremendously large number of ingredients. If we add into the mix our abilities to imagine and create new existences from each and every one of these ingredients, it is fair to say that the potential we have when we step into the ring of our practice is infinite.

So a grasp of what this infinity means, for ourselves, for our patients and for our practice, is really quite important for health practitioners. But have you ever seen it taught?

AMERY'S (VERY TRANSIENT) NO-MODEL MODEL

When trying to teach about concepts and realities that are hard to understand, teachers often use models, because models simplify and abstract out manageable chunks of information from the complex enormity of real-life.

In health practice education, we absolutely love models. In fact we love models so much that we have one chapter (Chapter 13) completely dedicated to them.

I don't want to be left out, so I am going to suggest, temporarily at least, a new model for health practice which takes account of the infinite emptiness of health practice.

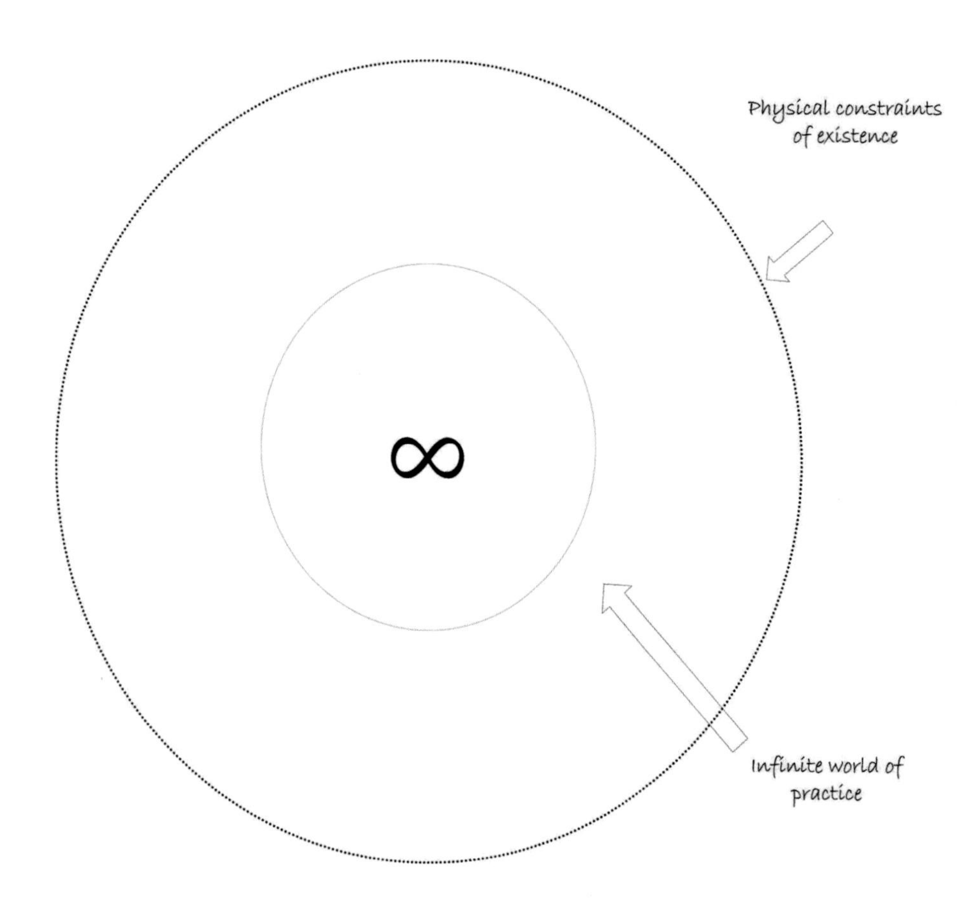

The Infinite Ring of Health Practice – aka Amery's No-Model Model

It won't last long. In fact it will be gone by the end of the book, so don't take it too seriously.

For now, though, my 'model' looks like this (*see* previous page).

It's not much of a model, in fact it's no model at all, but it does at least have the dubious merit of capturing the three important realities of health practice:

1. We are bounded: for example by our physical natures, by time, by space, by resources and by many other things.
2. Within those boundaries is nothing, emptiness.
3. From that emptiness comes infinite potential for creating better health.

At first sight, the combination of boundedness and infinity; nothing and something; may sound paradoxical.[15] But infinities can exist within the tightest boundaries, and the smallest space can be fully empty. It's all a matter of perspective.

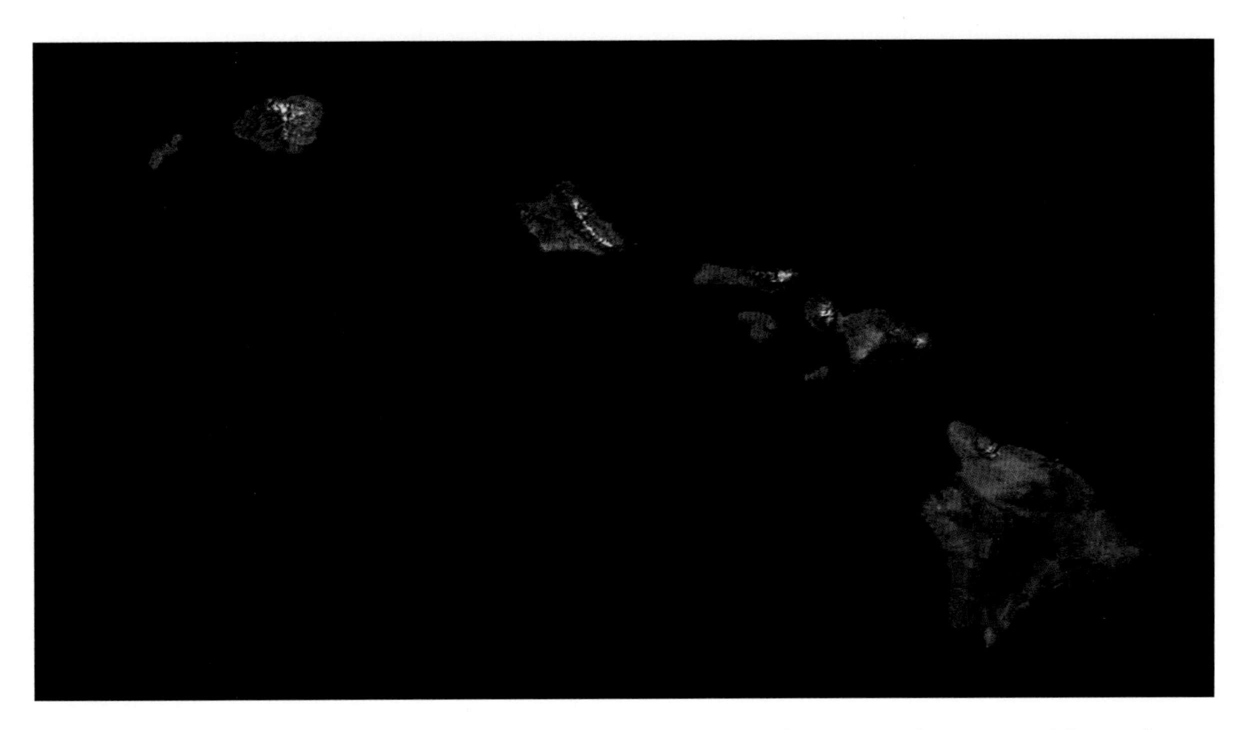

Infinite islands[16] – these islands may look small from space, but it's all a matter of perspective. If you measured them with a small enough ruler, on a small enough scale, they would be infinitely large.[17]

THE IMPORTANCE OF PRESENCE

If infinity and emptiness seem just too big and too unbounded to be useful to you in practice, try this concept instead: presence.

We can remember the past and we can imagine the future, but we can only actually be, and create, in the present moment. What is past is past, and we can do nothing about it. We can influence what happens in the future, but only by acting in the present.

The present, and our presence in the present, is therefore crucial for creativity. The nature of our presence makes a big difference to how well we create.[18] As we have seen, creativity requires imagination (divergence) and action (convergence). If we are fully present in the moment, we have a better chance of creating than if we are absent. If we are calm and focused, we have a better chance of creating than if we are anxious or distracted. If we feel confident and secure, we will have a better chance of creating than if we feel self-conscious or constrained.

So if we are to create effectively, we have to clear our canvas, to be mindfully aware of the present. Unfortunately, there are powerful factors that push against this in practice:

- Our internal censors: the sense that we 'ought' to behave, think and act in a certain way, according to the constructions and cultures of our professions, organisations and traditions.
- Our external censors: the objectives, protocols, policies, targets, guidelines and regulations that push us to apply particular approaches or solutions to diverse problems and individuals.
- Our internal inhibitors: factors such as tiredness, burnout, low mood, anxiety and sense of overload that make it extremely difficult to relax and diverge our thinking; and also reduce our energy and focus when we need to tenaciously converge and create.
- Our external inhibitors: time pressure, resource constraints, information overload, target-driven and safety-first organisational cultures, risk aversion and simple lack of opportunity to consider new ideas or new creations.

All of these things are really important, which is why we have gone through three whole workbooks just to get to this point. We have spent a lot of time looking at ways of recognising our censors, inhibitors and tyrants, turning them off, working round them, or even using them to our advantage.[19] But we can't do any of that unless we are mindfully aware of what is happening within us or around us.

To become mindfully aware we have to create space, from which we can take and maintain our perspective. If we remove all models, all we are left with is our existence, which is everything that we have.

Our existence is what we create, or co-create, in our consciousness. Within our consciousness we can imagine or create anything (and everything). Outside our consciousness, we cannot be said to exist in any meaningful way.

CREATION THROUGH IMAGINATION

Because we are conscious, we create each moment as we go along. Until we die, we can't stop creating, because creation is the fundamental property of consciousness. We can create our moments blindly, or we can create them with our eyes open, trying to engage with everything and everyone with whom we are currently in relationship.

When we create moments in our practice, we hope to create them compassionately, powerfully and skilfully, such that our creation is healthy. These moments of creation seek to pull together everything: the 'me', the 'we' and the 'other' through conscious and subconscious communication with ourselves, other people, and other entities.

Our brains and bodies are capable of dealing with vast amounts of different sorts of information, and pulling these together into a single, integrated moment of existence. But in order to actually 'live' in that moment, and actually try to create the next moment, it is very helpful to engage our imaginations.

We cannot create without imagination, because it is imagination that creates our experience of (and existence in) everything. But all this talk of imagination and dreaming might sound quite deranged, and hardly helpful in the hard-nosed, rough and tumble of actual day-to-day practice.

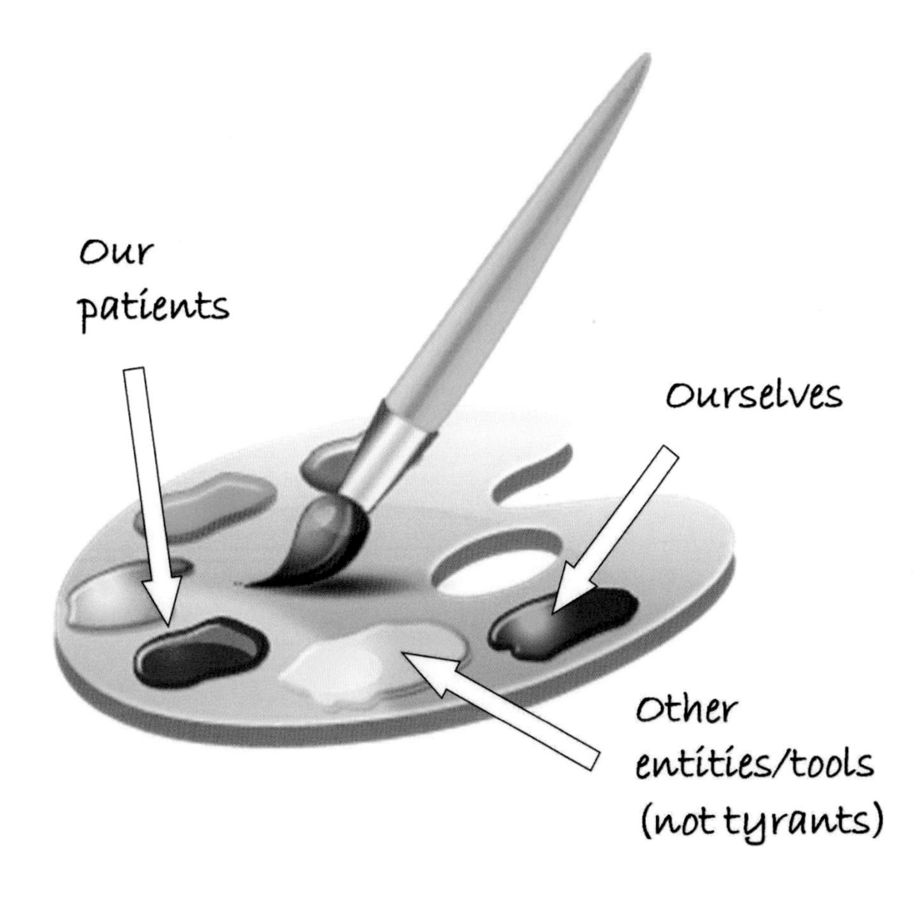

In practice, we can create infinite variations of 'better health' using just three primary colours: ourselves, our patients and our tools. But we need a blank canvas on which to paint: the infinite present.

Too right. It sounds like you were playing with those substances again.

If so, let us try to put that straight. As we will hopefully see, there is no contradiction between using our imaginations and being hard-nosed. In fact, as we saw in workbook 1 and again earlier in this book, the achievement of a flow state is extremely hard-nosed, and not easy. It involves marrying and balancing convergent and divergent thinking, beliefs and actions, mind and heart. Whether we create better health with a scalpel, with our words, or with our medications, we will do it faster, better and more effectively if we can find that balance.

Faster? Better? More effective? Very 21st century . . .

Being creative is not the same as being dreamy. Being aware of infinity is not the same as being empty-headed. Using emptiness as our canvas for creation in practice is practical and pragmatic. Combining imagination with thought and action to create better health is what we are here for. If we can do all this more speedily, more effectively and more happily, well, what's not to like?

Ananda

Even Shakyamuni could never tame Ananda
but Kashyapa kicked him out and tamed him.
Throw away all you know.
Throw away all you don't know.
Then and only then one star shines bright.

– Ko Un[20]

Activity 4.2: Emptiness and presence (30 minutes)

When you are next at work, try to become aware of 'gaps'.

When you start to look for them, you will find them everywhere: between words, between patients, between thoughts, between letters on the page, between breaths.

When you find one, briefly collapse yourself into it, as if you have just pushed the pause button on the remote control to your life. Enjoy, momentarily, the space to stretch and to breathe.

Recognise your own presence existing peacefully amid the noise and confusion.

Press the 'play' button and come back, but stay aware that you are still present, and that the emptiness is still there to be filled.

Section Two

Having a Go

Chapter 5

This is it

Activity 5.1: Having a go (30 minutes)

Whether you have read all the books, just this book, or even if you have read nothing up until this point, it doesn't matter.

Just start here.

Begin by trying to imagine yourself as an artist and your next consultation as your canvas. Your task is to create 'better health' with your patients. Imagine what 'better health' looks like. Does it look like what you currently create?

If not, what can you actually do to make your creation of better health more effective?

Try to be methodical and practical. Step by step, how can you integrate more things more effectively and more efficiently? How can you create even better health with the tools you have available, and within the constraints you cannot escape?

How would your new creation differ from your current one?

So here we are.

This is it.

This is truly it.

This is our practice.

We are where we are, and we have what we have. We have ourselves. We have our patients. We have our tools (and tyrants).

There are models, but there is no model for reality, and reality is where we practise. We have experience of many moments, but we have no experience of this moment, because it is unique. This moment is empty, but we have to fill it, together with our patients.

Our task is clear, to create better health.

Enough talking already! Could you please just get to the point and explain what we actually do?

Fair enough.

In the next few chapters, we will look at what we can actually do to make our practice more creative, more integrated, without losing the skills, experience and expertise we have built up so far; and without ignoring the harsh realities, boundaries and constraints of modern-day health practice.

The suggestions are generic and intended to be helpful whatever kind of practitioner you are, and however you practise. So it doesn't matter if you are a surgeon or a shiatsu healer; a nurse or a naturopath, a general practitioner or a gastroenterologist. We all work with the same raw materials: ourselves, our patients and our tools.

We all aim for the same goal: to integrate everything into one thing – better health.

So why don't we pull everything together, integrate it, and try it for real?

This is it.

Chapter 6
Clearing

Activity 6.1: Baggage check (1–2 hours)

For the next few days, keep a notebook handy. Just before you start with the next patient, note down some of the baggage you are carrying.

It might be emotional stuff. But it might be other things too: preconceptions about the patient, about how you 'ought' to practise, models or protocols, contractual targets or organisational guidelines, worries about the future, a sense of insecurity or complacency, time pressure, ill health. You name it. It could be anything.

When the day is over, go back over your notes. How much did your baggage weigh you down and slow you up? How much more efficient or effective would you have been if you could have gone into that consultation clear, focused and poised?

We practise moment by moment. We have no choice. The possibilities for what may happen in the next moment of our practice is infinite. Approaching infinity is not an easy thing to do. Although we might chafe against constraint, we are utterly emptied by infinity.

The infinite ring of our practice can therefore seem a chaotic, frightening and exhausting thing.

That may be why models are so enticing. They give us a sense of security in the face of the vastness of actuality. But if we approach infinity in a positive way, it can be a place of fascinating exploration, startling discovery and wonderful creation.

In the face of infinity, we have only three choices.

1 We can refuse to enter it.
2 We can decide to demarcate, focus on and explore only one part of it (shutting out the noisy neighbours clamouring on the other side of the fence).
3 Or we can recognise we are not going to be able to control and contain the infinite possibility of co-creation, stop trying, and plunge straight in.

All of these choices are rational, but the first is clearly not compatible with a career in health practice. All of us tend to do a bit of number 2. We have to have some boundaries otherwise our expertise would become as dilute as to be meaningless.

So, for example, as a family practitioner I would not expect patients to come to me for major surgery, or obstetrics, or psychoanalysis, or osteopathy. If they did, I would be little more help than the layperson.

But, the strange thing is, no matter how much we try to contain and control the infinite ring, it still stays infinite. I can shrink my field to become ever more specialised, but there is still boundless possibility within it.

Which means, if we don't want to do number 1, we are stuck with number 3.

CLEARING THE RING

Before we start the next moment, it is helpful to clear the last.

It is very difficult to undertake any journey when our destination is unclear, the way ahead overgrown, or full of dragons, or when our vehicle is clapped out.

In the same way, when our health practice becomes cluttered, when we succumb to fear about impending challenges or exhaustion, when we lose sight of our values, when our minds become clouded, when we stop communicating, or when we act carelessly or thoughtlessly, we make it very difficult for ourselves to practise effectively or healthily.

But, at these moments we can remind ourselves that the infinite ring of our health practice is our very own creation. If we want our practice to be a clean creation, it helps to have the ring clear.

So let's clear it.

GETTING PERSPECTIVE

When we take an overarching perspective, the view is not always that beautiful.

You never get quite the same perspective on life as when you are standing on a high diving board contemplating going over the edge . . .

Our patients are going to die. We are going to die. We are probably all going to suffer en route. And there's nothing anyone can do to stop it.

We can choose to be nihilists. Nihilism is an eminently sensible and pragmatic perspective. Some of my favourite internal voices are nihilist.

Heaven
Wherever they've
gone they're
not here
anymore
and all
they stood
for
is empty
also

– Robert Creeley[21]

The way ahead may be hard. It might be frightening. It will almost certainly be tiring and may even psychologically or physically harm us. So we can choose not to do it.

> Not doing something that might harm us is also an eminently sensible and pragmatic perspective. Some of my favourite internal voices are really, really frightened.

On the other hand, let's take a different perspective. We are almost miraculously lucky to have been born, and even more miraculously lucky to have been born sentient, and still more miraculously lucky that we have been granted sufficient capabilities to become health practitioners.

As sensible, rational people, we know that any relationship or co-creation is constrained in many, many ways. We can rage against that, but it won't change much. Constraints are part of existence.

If not, we'd all have wings.

But, while we have big and fixed constraints about us, we do have an awful lot of potential to make the most of this incredible thing called existence, and use our work as a tool to explore and create wonderful new opportunities.

> And if we are even luckier we can get paid for it.

So, despite all our warning voices, many people decide against nihilism and terror, and choose wilful optimism. We can see the glass as being half-full rather than half-empty. If we'd been given the choice between existing and not existing, we know we would choose.

But being optimistic is not the same as being stupid. We have to be honest, with ourselves and with our patients. We can't do much, but we can do something.

If we step into relationships with our patients with due value accorded to compassion, honesty and courage, almost anything we do will likely be helpful. Just sitting together with someone who cares is a big boost for all of us.

So, we are already ahead on points.

> *It ain't what you do it's the way that you do it*
> *(and that's what gets things done).*
>
> – Melvin Oliver and James 'Trummy' Young[22]

HEALTH PRACTICE AS 'BEING' NOT 'TRYING'

This chapter has been deliberately light-hearted, and evidence-light. However, there is a strong assertion behind it. In fact it is an important as assertion of the whole series of workbooks. It is this.

Unless we connect with our patients, and with all the other entities of our practice, we cannot properly communicate with them. Without accurate communication, we have inadequate information. Without information, we handicap our consciousness, and prevent it from creating a present that gives us a clear and full insight into whatever is happening in the moment. Without insight, into our patients, our practice and ourselves, we are practising blind.

To fully connect, we need to be fully aware, without distraction. That means 'being' in the moment, rather than 'trying' to achieve tasks or goals, or being distracted by emotions, thoughts and feelings. These things distort our perspective, and so we lose clarity.

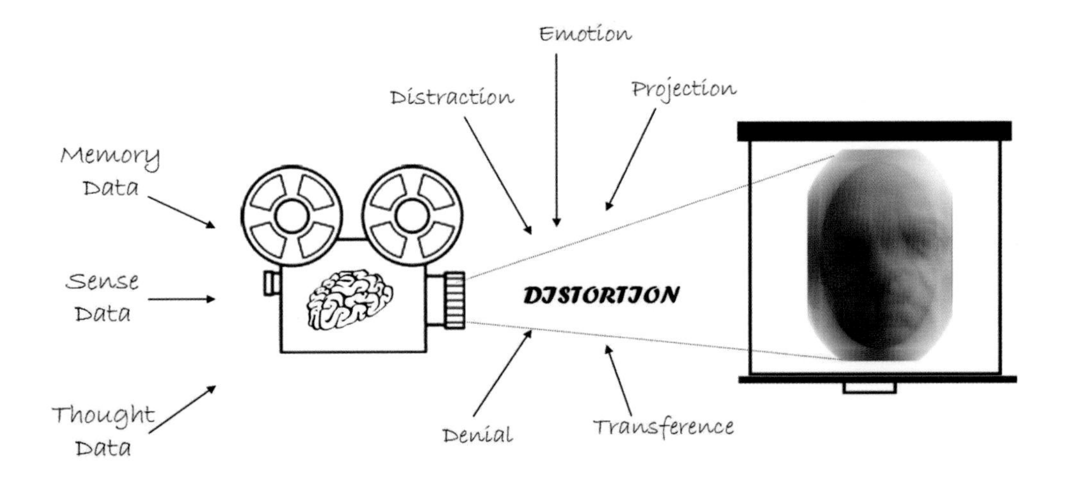

It's not that targets, money, tiredness, information, environment, family and countless other things are not important; or not relevant; or not useful. It is just that, for those few minutes where we 'merge' with another sentient being, the more focused and connected we are, the more effective we will be, and considerably more efficient too.

Clearing one's mind takes a bit of practice, but it is not difficult to do. There are numerous ways of doing it: progressive muscular relaxation, prayerful contemplation, stillness meditation, the 'whoosh' technique, etc.[23] We covered some of them in workbook 1, and there are many others around. With a little practice, we can 'telescope' the whole process into a moment, so that we can clear our minds within a couple of seconds.

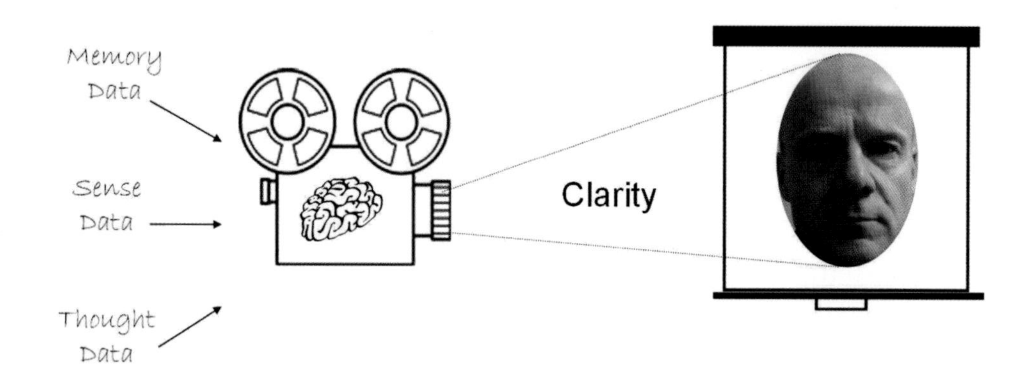

Creating an absence so that our patients can create a presence, finding that still point in the middle of our internal noise and clutter, clearing our minds, clearing the ring: these are core skills for practitioners.

They are core skills that we may not (yet) find on our training syllabus, but they are ones we should practise, practise, practise.

But always expecting to fail, and always smiling gently at our own failures.

Parting Nothing

Being still,
To fill space
I empty
The place
I was.
To be present
I become absent

– JA

Activity 6.2: Clearing the infinite ring (5 minutes, but getting quicker with practice!)

Remember the lessons of workbook 1. Choose a method of clearing your mind that suits you and practise it so that you are able to do it quickly, within a few seconds.

The aim is to be able to do this between each patient.

For example . . .

Sit or stand wherever you normally practise. Close your eyes if you can; briefly look down if you can't.

Take a deep breath all the way in, then allow it to release, slowly and gently.

Quickly visualise all the of bits of tension in your body: jaw, forehead, neck, shoulders, back, tummy, pelvis, buttocks and feet. Now visualise yourself under an incredibly invigorating shower, any temperature you want. Mentally turn the tap on and feel the powerful gush of water washing right down through your body, rinsing and draining the tension out.

Don't worry if it's not completely clear. Good enough is good enough.

Now turn your inner eye to your mind. Visualise it as a huge chamber full of noise and activity: thoughts, feelings, emotions, ideas and concerns. At the side of you is a button. Push it and watch as the roof suddenly disappears and the floor drops away into nothingness. Watch all the noise and clutter plummet until, after a second or two, they are gone.

Again, don't worry if there are a few stragglers. Smile, wave and then ignore them.

Feel the emptiness and potential of the moment.

Make a mental note to commit yourself to whatever is coming next, then open your eyes.

You are about to jump.

Chapter 7

Awakening

Activity 7.1: Touching the void (variable)

Go to a swimming pool, or a lake, or the sea. You are going to jump in. It's better if it's a bit cold or scary or challenging, but it doesn't have to be.

As you move towards the jump, keep your mind's eye fixed on itself. Watch what is happening in your head: your thoughts, your emotions, your feelings. Get closer to the point of jumping until you are there, poised to go.

Clear your mind as much as you can, but keep the internal focus fixed in your mind's eye. Then jump. As you do, notice that there is the briefest thought followed by an even briefer moment.

Notice that, after wavering for a while, you finally commit to the jump. The commitment leads to a 'nothing' moment, where you are neither jumping nor not jumping.

Then experience the rush as gravity takes over and you plummet to the water below.

When it comes down to actual health practice, this perspective-taking and clearing, very brief though they are, are not actually going to get anything done. We are busy, and have to get on.

And that means we have to jump in.

COMMITTING

To stretch the metaphor a little further, we can stand on the edge all day, pacing back and forth, worrying and fretting. But, however much time or effort we spend, we still only have two choices: the humiliating climb back down the ladder; or the terrifying plummet.

At some point, whatever action we decide to take, we have to ***commit***.

In health practice we commit in a fraction of a second, and we do it so often that we mostly are not aware we are doing it, but we surely are, and it is a crucial moment, because it is at this moment that we decide, consciously and subconsciously, to commit ourselves to whatever is coming next.

If we commit only half-heartedly, the jump will not be as streamlined, the dive not as pure. If we commit, if we can close all else out of our mind, and if we commit all of ourselves to the next moment, into the creation we are about to create, we give it (and ourselves) the best chance of effective, skilful action.

AN EXAMPLE

It's 07.45 and I am alone at my desk in my consulting room. I have had a look at the patient list, a read through the records. I have 25 patients to see, including two of my 'heart-sinks'. I have taken two or three interruptions already, and work hasn't even started. We haven't yet hit our targets for the management of various chronic diseases; and points equals prizes. I need the money as we have a holiday coming up, and my big family is ferociously expensive. I have my appraisal next week, and I haven't completed all my paperwork, let alone done all the things I said I would do this time last year. I'm fairly p***d off as there was a big family argument before I left home and I stepped in some dog-s*** on the way out, I have a sore head (self-induced) and I really badly need a coffee.

So I get a coffee (at least that's one thing I can do).

Then I try to relax my posture and look just to the side of my computer screen, where I have pasted a tiny piece of paper, with the word 'light' on it. I close my eyes briefly and see myself under a big circus tent, on a high diving board above an impossibly tiny pool in circus ring, in which are all the tyrants of my practice and my existence: bustling, jostling and looking up at me.

They are all there: the 'me' tyrants (my dodgy values, unhelpful beliefs, frightening memories, labile emotions and bad habits); the 'we' tyrants (my heart-sink patients, the unhelpful drama triangles and narratives I get drawn into) and all the various 'other' factors (the targets, guidelines and tick boxes, the computers, the rules and regulations, annoying colleagues). I nod and wave at them, but studiously don't engage.

Then in my imagination I turn to my left and see a big green button saying 'push me'. So I do.

The floor of the ring collapses, and the tyrants plummet. I look up, the roof of the big top starts folding, again and again and again, becoming smaller and smaller until – ping – it disappears, revealing a beautiful, blue, cloudless sky. I look down to find the small pool in the ring has been replaced with a vast, clear, blue, waveless sea.

I commit to jump.

With a deep sense of hope and peace, I step out, and am amazed to find I can walk on air (I still like to hold on to a bit of my messiah complex), until reality hits and I plummet headlong into the mad, crazy circus of health practice.

Constantly risking absurdity

Constantly risking absurdity
and death
whenever he performs
above the heads
of his audience
the poet like an acrobat
climbs on rime
to a high wire of his own making
and balancing on eyebeams
above a sea of faces
paces his way
to the other side of the day
performing entrachats
and sleight-of-foot tricks
and other high theatrics
and all without mistaking
any thing
for what it may not be
For he's the super realist
who must perforce perceive
taut truth

before the taking of each stance or step
in his supposed advance
toward that still higher perch
where Beauty stands and waits
with gravity
to start her death-defying leap
And he
a little charleychaplin man
who may or may not catch
her fair eternal form
spreadeagled in the empty air
of existence

– Lawrence Ferlinghetti[24]

DISSOLVING

Just after committing, and just before jumping, there is a fractional, fractal moment, which is neither jumping, nor not jumping. Just like Schrodinger's cat, or like the quantum wave-particle, we are simultaneously in two states of existence: jumping and not jumping.

This is an infinite moment of pure emptiness. It is neither doing nor not doing, acting nor not acting. In this tiny moment, we touch the void, the point of stillness at the centre of our spinning wheel.

It is the point where we dissolve as anything at all, becoming one with the nothing out of which everything exists.

This is the Zen moment of health practice. It is the moment when time stops, when we are neither with the last creation nor with the next creation. We just are – part of the pregnant nothingness. It is a moment of pure potential, out of which the whole of the rest of our life will be created.

Awareness of this still point grounds us before we plunge in. It is therefore a hugely important point. The amazing thing is that it is always just in reach, ready to help us find instant perspective and grounding, if we can, for a millisecond, just allow ourselves and our egos to dissolve.

And yet we so rarely acknowledge it, let alone accept it for what it is: one of the most useful tools in our practice (or indeed in our lives).

JUMPING

Now it's starting to get interesting, because this is where all the hard work we have done in workbooks 1, 2 and 3 starts to come together. We are nearly at the moment of truth. The crowds are watching, our heart is racing, and our mind has whited out, because we are about to jump.

This is the point of maximum danger. This is the point where all our tyrants can invade and overthrow us, by full frontal assault or sneakily round the back. Within milliseconds we can lose our focus, get drawn into internal distractions and worry, falling into that same old trance.

That's no good, because to practise effectively and efficiently, we need to be awake.

And there is nothing like a good plummet to wake us up.

You never feel quite as awake as when you are plummeting through a clear blue sky.

WAKING UP (FLYING, NOT FALLING)

Let's start by becoming awake, fully aware, fully mindful.

When we start to practise, each morning, each patient, each moment, it can feel like a crazy, out-of-control, spreadeagling, cartwheeling plummet. We can feel as if we are buffeted by winds and currents over which we have no control.

As soon as one moment finishes with us, the next one whips us up and off, who knows where?

So it's really, really useful to remember that time is relative. The speed that the physical world operates at is, for example, quite different to the speed our subconscious works at. We can dream whole years in the space of a few minutes.

If we can slow things down a bit, instead of plummeting, we may be able to do a little gliding, or perhaps even flying.

So, in analysing those first, vital few milliseconds of our jump, of the moment we start to practise, we are going to slow the clock right down. It's not crazy. Our subconscious can operate at a much faster pace than our conscious mind, so we can cram in a huge amount, millisecond by millisecond.

> **Did you know, for example, that our brain seems to know it is about to solve a problem, several milliseconds before it actually solves it?[25]**

Imagine leaping into the infinite ring of your own practice. Catch that image and freeze-frame it in your mind. Then, frame by frame, millisecond by millisecond, we will start to advance, and watch as your leap evolves.

As we do it, don't think. Just watch. Your subconscious mind will notice all of this, and do the choosing for you, without any effort, and incredibly efficiently, as long as you don't interrupt it. Thinking is a trap at this stage, and it will slow you down, as your conscious mind is not meant for noticing; it's meant for analysing, sifting and calculating.

Imagine your practice captured in old-fashioned cine-film, frame by frame.

Look at frame one. Where are you? Is it a consulting room, or a ward, or an operating theatre?

Move to frame two. Become aware of all the 'me', 'we' and 'other' entities that are at hand, but don't get distracted by the dizzying number or bewildering variety. Remember they are tools, not tyrants. Each moment has a particular need, so you only need those tools that suit this particular moment.

Allow your subconscious mind to shuffle these people around instantly, so those tools that are useful are at the front and those you don't need can stay drifting in the background.

If the noisy ones at the back protest, either let them slip forward or ignore them. Their time will come.

Move to frame three. Look up, at your (very big) mental clock. It's there, right above your head. It's not ticking, at least not yet.

Notice there are three marks on the dial. The first is the amount of time you'd like to take. The second is the maximum time you can take without putting out other patients or other people, or without putting too much strain on your resources. The third is for 'emergencies only'.

This 'for emergencies' mark is a clever mark, as it can float, taking just the right amount of time, and pushing other bits of time aside. We don't often need to use it, but when we do, we will find other resources we didn't realise we had.

In a moment, we are going to start the clock, and, when it's started, we are going to check it occasionally.

Move to frame four. Imagine the space inside your head.

What is in your mind? What are you thinking or worrying about? Don't engage with the thoughts, just quietly observe and name them. Watch them floating about like clouds in the sky. Decide which ones you want to hold onto and let the others drift off past the horizon.

Don't worry; they'll come back soon enough.

Move to frame five. Look in a mental mirror. Look at yourself, as if you were a stranger seeing you for the first time. Who do you see? What signals are you communicating? Do you need to quieten your messages down, or are they peaceful enough to allow the patient space to express herself?

She'll take less time and do it more effectively if she doesn't feel distracted by you.

Move to frame six. Notice that the ring is not empty. It has one very important person floating right above the middle of it. You.

Move to frame seven. Look up. Imagine your gods (if you have them) or your beliefs and values (if you don't) shining above you like suns in the sky. Allow yourself to absorb them, and feel yourself being directed and focused by them.

Move to frame eight. Look down, at the deep pool way beneath you. It's labelled 'my resources'. Have a look inside and see what resources are available to you at this moment. It may be very full or quite empty. Either way, you are going to hit it, so you may as well make the entry as smooth and painless as possible.

It doesn't matter. You can only give what you can give.

Make a bargain with yourself. This is what I can offer right now. If this crazy circus demands more, I'll either get help or I'll reschedule.

But right now, I am clear, I am awake, and I am flying (not falling).

It's time to start my practice.

Awake!
Gazing at the finger pointing to the moon.
Wondering at the sound of the call to prayer.
Searching for patterns amongst the tears.

Dreaming on,
Hypnotised by thoughts,
Addled by ideas,
Cushioned by concepts.

Starving before a feast-laden table.

Awake!

– JA

Activity 7.2: Freeze-framing (really, really fast, but incredibly slowly)

For the next few days, practise freeze-framing.

The moment before something starts; perhaps the moment before a patient walks in, or before you first smile, or before you first cut the skin – freeze-frame the moment in your mind.

In that moment, allow all tension, distraction and concern to drop away. Watch the space around you to open up, and become fully awake to everything, allowing and trusting your subconscious to provide you with the tools that the moment needs, so that you have absolute focus and absolute clarity on the patient and the matter in hand.

Or at least a bit more focus and clarity than usual.

Chapter 8
Connecting

Activity 8.1: Your first contact (15 minutes)

Remember something from your practice that is clearly etched in your mind. It may be anything, perhaps the very first time you were let loose on a patient, or the very first time you saw someone born, or the first patient you saw die.

Remember the details. What was he wearing? What did she say? What did you say? What were the problems? How did you do? What was going through your mind?

What was it about that patient that sticks so vividly in your mind? What made you so open?

Do you still have that openness in your practice now?

THE GOLDEN MINUTE

Contrary to what we have come to believe, time is a flexible concept and a flexible entity.

This is particularly useful as, when we first make contact with the patient, a tremendous amount happens in the first few seconds, and time seems to fly. Fortunately, our subconscious is capable of slowing down the clock and taking in a huge amount more than we are conscious of, as long as we don't close down our subconscious with thoughts, or worries, or plans, or distractions.

If we miss what happens in these first few seconds after contact with our patient, or if we fail to make use of that, we will have to play catch-up for the rest of the consultation. That means we will work more slowly, less efficiently and less effectively.

If our minds are not clear, we cannot be mindfully aware. If we are not mindfully aware, it is impossible to make the most of this instant. If we are thinking about what we are going to say, or the targets we have to hit, or the coffee we never got, we will be distracted and will miss valuable clues, and make our job much harder and more laborious.

INTRODUCING THE PATIENT INTO THE INFINITE RING

The good news is, in these first few seconds, we don't have to think convergently. Convergent thinking is for analysing. Right now all we want to do is to notice, to diverge, to be aware, to be mindful. We are eagles at 1000 feet, but with such sharp eyesight we can see every little detail.

And we are flying, not trying.

As we saw in workbook 3, when we meet someone, we take in, instantly and automatically, many, many details about the situation. This is presumably evolutionarily important. Usually it happens unconsciously, but the details remain with us and accessible to us.[26]

'Zen mind, beginner's mind'[27]

However, we will take in much more, and notice much more, if our mind is still and our thoughts are quiet. With practice, we can become more conscious of the information about our patients we are taking in during the first few seconds: how they dress, how they walk, how they move, how they talk.

Clue, after clue, after clue.

This bit doesn't require any technical training. In fact it requires 'un-training'. Here we need 'beginner's mind'.

As sentient human beings, we are incredibly adept and effective at communicating instantly with our fellow sentient beings. It is completely innate and automated. We can choose to cloud our awareness of that communication with preconceptions, worries or plans. Or we can choose to look at it with the eyes of a child, a beginner, seeing for the first time. Then we can really see.

If you doubt it, think of those people who make you feel happy, or sad, or cross, almost the instant you meet them.

Or think of a kiss.

This connection is instantaneous, and rarely does it lie.

THE FLOOD

Once contact is made, all of a sudden the present kicks in. A flood of visual, auditory, olfactory and sensory information hits from the right. Another flood of thoughts, ideas, worries, memories hits from the left.

Our circus ring is suddenly crowded with storytellers, acrobats, clowns, jugglers, plate spinners, human cannonballs, dancing bears, roaring lions, charging horses, high wire artists, Uncle Tom Cobbley and all. What's worse, you are all of them, and none of them. It's chaos. You're chaos. The crowd is on its feet . . .

Breathe in. Breathe out. Slowly.

All the way out.

And a bit more.

Breathe in.

Look up.

If we can withstand the initial flood, keep our bearings, and keep aware, we will soon remember that we are in control. We are the ring-masters. Our initial job is to guide, to direct, to contain. It is not to become part of the show.

Oh alright. I know. It's the best I could do.

'Whaam!' –Roy Lichtenstein[28]

DEEP LISTENING (AND WATCHING)

When we make contact with our patients, our first job is to listen – with our ears, but also with our eyes, our minds and our intuitions. We dedicated a whole workbook to the art of listening, communicating and co-creating with patients. But perhaps, out of that whole workbook, the most important skill for effective and efficient practice is deep listening.

It's extremely tempting to speak when we first meet someone, particularly when we are on home turf, and they are a bit anxious. But the best way to save face is to keep the lower half shut. When we are in transmit mode, we don't hear. What is worse, if we start speaking before our patients do, we can wipe out a whole raft of useful clues.[29]

DON'T INTERRUPT. REALLY DON'T. REALLY. DON'T.

It is worth reminding ourselves of times where we may have been patients rather than the practitioner. Before we go in to see the practitioner, we go through in our minds what it is we are really worried about, what we are going to say, what we think

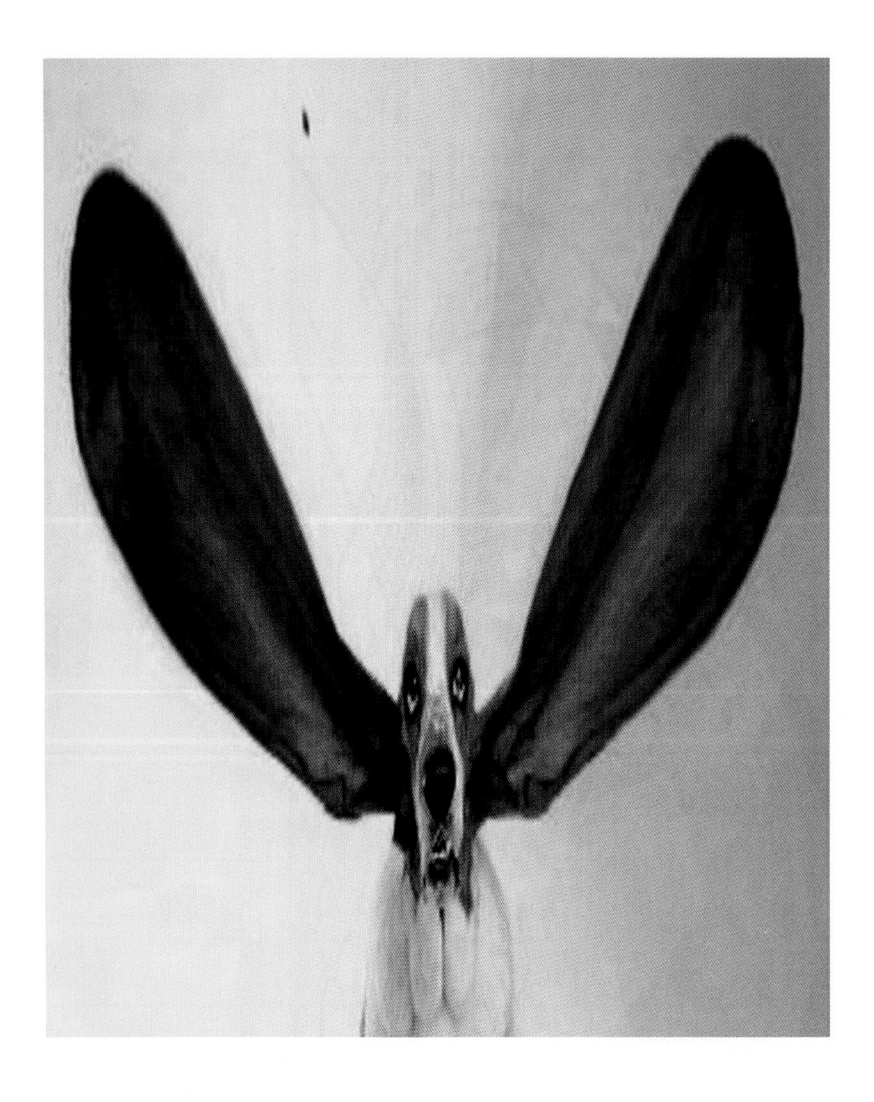

Remember me?

might happen. On the other hand, we are a bit frightened, off our territory, and so easily put off.

Practitioners are often worried about time-keeping, so we may try to set things off with an opening phrase like 'What can I do for you today?', or 'How's your back?'

Here be dragons. Even those few words can put patients off.

In those first few seconds we need to be mindfully aware and quiet. If we listen, we will hear so much about the patient: the content of the words, the delivery of the words, the gaps, what is left unsaid, what is hinted at. Stopping this flow is the equivalent of damming the river with which we wish to merge.

Our time to perform will come, just not yet. The clock is ticking, but we are probably still only one or two minutes in. There's plenty of time. Listening deeply is extremely efficient, because it will save us from disappearing up all sorts of blind alleys and closed loops later on.

THE VERY USEFUL MIRACLE

In workbook 2 we looked at the ways that we can connect and co-create with our patients, and this is where it all comes to fruition. It turns out that, apart from being nigh-on miraculous, our ability as conscious beings to co-create our existence is an incredibly useful tool, because it tells us more in a few instants than we could ever learn in hours of simple questioning and examining.

If we can keep our analytical minds still and quiet when we initially connect with patients, what we gradually become aware of is that we are becoming aware of what it is to be the patient. Not as an outsider looking in, but actually experiencing the patient's existence.

It is difficult to overstate the usefulness of this: our ability to connect with incredibly 'thin slices' of information. If we can experience what our patient is experiencing, we can find out much more about their problems, their ideas, their concerns, their hopes, and their wishes than we could hope to get from the simple 'history and examination' of traditional practice. What is more, we can get a much better idea of what might work, what would work best, how much they understand, how much they buy into the plan, and how much support they need.

How? We don't know, although in workbook 2 we looked at some theories and evidence. But, in simple terms, it is presumably because our subconscious mind is so much quicker, effective and efficient at connecting and reading other people than our conscious mind.

Of course we need to engage our conscious mind too. Our impression of the existence of others will always be clouded, and coloured. So we need to check our ideas, safety net our diagnosis, and overtly negotiate plans. But these come second.

First, we need to give our subconscious time and space to do its work.

Here we are all, by day
Here we are all, by day. By night, we're hurled
By dreams, each one, into a several world

– Robert Herrick

Activity 8.2: Practising mindful contact (1 hour)

For the next few sessions that you are in practice, have a go at allowing the golden minute to absorb you, and you to absorb it.

Before the patient comes in, or before you go to the patient, consciously clear your mind and emotions using some of the techniques we have been through. Imagine you are a child, or a new student, seeing things for the very first time.

Once contact is made, go into deep listening mode, allowing all the information to flood over and through you. If you feel anxiety arising, or feel the urge to interrupt or intervene, or notice your mind shooting off down a particular path, gently bring it all back to the present, and wait.

Allow your subconscious time and space to do its job. Trust it. It will work far more efficiently than your conscious mind. As you do, become aware of feelings and thoughts arising in you that were not there a minute ago. Recognise the possibility that they belong to you, but also recognise the probability that they come from your patient.

Allow yourself to 'feel' the diagnosis, to 'intuit' the problems, and the answers to reveal themselves to you. Trust what comes. Don't interrupt.

When nothing else emerges from your patient, re-engage your conscious mind, your rationality and your practicality. Take nothing for granted, safety-net everything you need to ensure you will do no harm, and check all your plans with the patient.

See how much time you save, and how much more effectively you work.

Chapter 9

Trust your intuition (but check it too)

Activity 9.1: Use your intuition (1 hour)

Intuition seems to have gone a bit out of favour in health practice, which is a shame as, used well, it can be both fast and effective. Test it and see its strengths and weaknesses for yourself.

First (unless horse racing is familiar to you), pick up the racing pages of a paper and 'intuitively' choose the most likely winners. At the end of the day, count how many you get right.

Next day, do the same, but introduce some threat. Put some money on it, enough to hurt if you lose it. Notice how you start checking the form and reasoning.

Third, try it in practice. Give yourself a minute after the patient has walked in the door, and intuit what the likely diagnosis or problem is. See how many you get right or nearly right.

Consider why your intuition works better in your practice than in an area where you are a novice.

Gradually the flood subsides, and we find ourselves in possession of enormous amounts of information: from the patient, from the notes, from the computer, from our memories.

Now it's time to create something. It's time to think. But what is thinking? The problem is we aren't entirely sure.[30]

Definition: thought – that which one thinks

We don't have to be Socrates to work out that one's a bit circular.

THINKING: FAST AND SLOW

To create anything we have two tasks: to come up with an idea and then to put that idea into practice. If we only do the former, we will be hopeless dreamers. If we only do the latter, we will be unconstructive pedants. Neither makes a particularly effective practitioner.

So, if we want to co-create better health with our patients, we need our thinking both to 'diverge' (looking for new connections and solutions) and also to 'converge' (checking ideas for feasibility, safety and efficacy as well as planning and executing the ideas). As we saw in Chapter 3, divergent thinking (sometimes called intuition) is much quicker and less effort than convergent thinking (sometimes called reasoning), but is more liable to bias, so needs to be checked to be safe.[31]

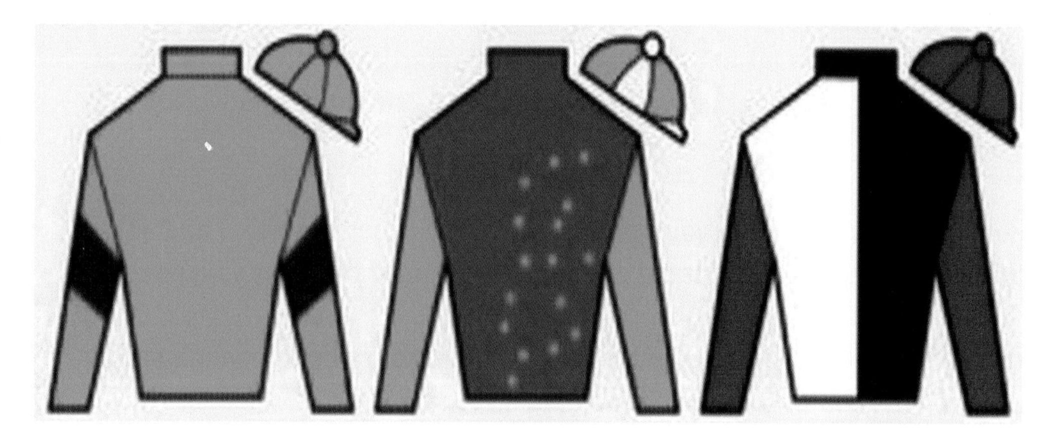

Intuition and expertise – test one. The jockeys' colours above were worn in a recent horse race. Use your intuition to tell you which horse won.[32]

THE PRACTICAL APPLICATION OF CONVERGENT AND DIVERGENT THINKING

In practice, we can easily see that there are strengths and weaknesses of both converging and diverging.

When we are connecting with our patients, listening to their stories, trying to understand their ideas and concerns, we need to be divergent. This enables us to be open, aware, and to make rapid connections and associations. If we jump to convergence too soon, we will close down and slow down, missing important prompts and cues, heading off down blind alleys, becoming too defensive too early and making our practice less efficient.

The watchword for overuse of convergent thinking is 'paralysis by analysis'.

When we are safety-netting for dangerous problems, sieving for localising or diagnostic symptoms, carrying out a methodical examination, or planning treatments, we need to be focused and convergent. If we stay in divergent mode during these times we are more likely to make dangerous errors of judgement and mistakes, be distracted, over-optimistic and risky.

The watchword for overuse of divergent thinking is 'happy complacency'.

But the news is not bad. In fact it is trebly good.

First, as we become more experienced and skilled, tasks that initially require the effort and time of convergent thinking become more automated and intuitive. Second, as we become more experienced and skilled, we can trust our intuitions far more.[33] Third, we always have the opportunity to double-check our intuitive answer for logic and safety.

Intuition and expertise – test two. Look at these pictures for a very brief moment, say 2–3 seconds. Use your intuition to decide which one is having a heart attack. Can you 'intuit' the other ones?[34]

PROBLEMS WITH DIVERGENT AND CONVERGENT THINKING

Have a think about this scenario and question:

Barry has just been admitted to hospital with a heart attack. Which is more likely, that Barry is a smoker or that Barry is an obese, diabetic smoker?

The first answer to pop into your head is probably the latter (he is an obese, diabetic smoker). If so, that's wrong. The number of smokers is greater than the number of diabetic, obese smokers, as the latter is a subset of the former. So he is more likely to be a smoker only. The reason we tend to think he is more likely to be a diabetic, obese smoker is that we have already created associations between heart attacks, smoking and obesity in our mind; we forgotten to check the logic of those associations. That's divergent thinking at work.

So, because divergent thinking is quicker and much less effort, we can jump to it without 'checking' using our logical reasoning, and make significant and serious mistakes if we over-depend on it in practice.

On the other hand, let's imagine Barry comes to see us with chest pain before he has his heart attack. If we see a sweaty, obese diabetic sitting in front of us holding his clenched fist in front of his sternum as he describes his pain, we will be on the phone getting him to the cardiac unit within a minute.

If, in this situation, we refuse to accept our intuition and insist on taking a thorough, complete history, carrying out a complete physical examination, reading up the statistical probabilities of various differential diagnoses and working through the logical algorithms of possible management before we come to a decision, Barry may well have expired before we do anything useful. That's convergent thinking at work.

What's more, because convergent thinking draws us 'inside' it tends to blind us to what is going on around us, so Barry's sad demise may have completely escaped our notice until we lift our head up from the guidelines.

So, because convergent thinking is slower, more effortful and more all-absorbing, it can slow us down, exhaust us and blind us if we over-depend on it in practice.

FACTORS THAT MAY AFFECT THE WAY WE THINK

We might like to think that we are in charge of how we think. If so, we might like to think again.

For a start, we tend to have a natural 'preference' for either convergence or divergence (although all of us can and do practise both).

There are also many factors that can push us towards, or away from, and bias us in a particular type of thinking.[35] The main factor is the degree of 'cognitive ease' we feel.

As we saw in Chapter 3, factors that increase our sense of 'cognitive ease'[36] will make us more likely to think divergently, whereas factors that increase our sense of 'cognitive threat' make us more likely to think convergently, but there are other factors too.

It's no surprise we often get our best ideas in the bath or shower.

There is no evidence for this that I am aware of, but it seems to me that, in modern health practice, there are far more primers of convergent thinking than of divergent thinking. For example, time pressure, burnout, exhaustion, resource pressure, targets, guidelines, busy and threatening environments, information overload, rules and regulations all make us feel cognitively ill at ease.

This also fits with my experience of teaching and training consultation skills. Learners more commonly run into problems being too convergent (particularly 'paralysis by analysis' and becoming 'closed down' to the patient) than in being too intuitive.

The point is, we sometimes forget how fast and effective intuition can be. Given that we use it in scenarios where we are experienced and have expertise, it is pretty safe and effective too. As intuition is both much faster and much easier than reasoning, it seems a shame if we are nervous about using it.

EBBING AND FLOWING

Practice ebbs and flows. There is a time for listening and a time for acting. There is a time for perceiving and there is a time for judging. In the same way, our thinking ebbs and flows, sometimes convergently and sometimes divergently.

If we can be aware of these ebbs and flows, we can try to match our thinking to the moment, being open and intuitive when listening, searching and generating, and convergent when checking, sieving and planning.

However, in order to become aware of our thinking in practice, we need to become aware of ourselves, our challenges, our patients, our environment, and the other 'primers' of practice too. For example, we might want to become more aware of the following.

- Myself: what is my preferred style? Am I high or low at the moment? Do I feel stressed or calm? Do I feel like I need some coffee and chocolate; or a beer?
- The challenge: is the current challenge to maximise safety or to maximise ideas? Is there risk to either me or my patient?
- The patient: is she someone who seems happier with chatting and exploring, or someone looking for concrete answers and solutions?
- The environment: is it busy, noisy and stressful? Or is it peaceful, quiet and relaxing?
- Other primers: how are issues, questions and agendas being put to me: in an open or closed way? Am I being pushed into following guidance, obeying regulations or chasing targets; or do I have a free hand to approach this how I think best?

It helps to have quick tricks to switch from one form of thinking to another. For example, we can use rapid relaxation techniques. We can sit forward or backward in the chair. We can place different visual triggers to relax or to beware in our line of sight (like pictures, images or words). We can make sure we use, or avoid, stimulants like coffee and chocolate.

It can also help to verbally 'signpost', both to ourselves and to our patients, when the focus of the consultation needs to switch from convergence to divergence, or vice versa, as in the following examples.

- Let me just check what you are saying. (divergence to convergence)
- So you have come about your pain, but I am not sure I have quite got the whole picture. Please can you expand a bit more? (convergence to divergence)
- Well, I think I can see how it must look to you. Can I quickly run through some questions to rule out anything significant? (divergence to convergence)
- Thank you for painting that picture of yourself. It was very helpful. I think it is time for me to examine you now. (divergence to convergence)
- Well, the examination was normal. Why don't we spend a little time reflecting on how we might manage things? (convergence to divergence)
- OK, those are all possibilities, but let's now think of some concrete plans. (divergence to convergence)

INTEGRATING AND HARMONICALLY BALANCING OUR THINKING IN PRACTICE

As we go about our practice, as we go about creating better health, we don't want to be complacent or paralysed by anxiety, but we do want to be quick, efficient and effective. That means being able to recognise and use the most effective form of thinking at the right time.

The key here is the word 'recognise'. As with all tools, we will use our thinking more effectively if we are **doubly aware**, not just of what is happening with the patient, but also of what is happening in our own minds. When it comes to thinking, this is hard, because it means that our mind has to multitask. We need sometimes to diverge, and other times to converge. We need to be aware that we are converging or diverging, and we need to make judgements on which type of thinking is most likely to be effective in the particular situation. We need to ebb and flow, each type of thinking in seamless interaction with each other.

And we need to do all this neutrally, without intuiting or reasoning about the intuitions or reasons.

That's an awful lot of thinking!

Paradoxically (and you just have to love paradox), the secret about managing all that thinking is not thinking about it. Just like when we are driving, or when we brush our teeth. We hand control over to the part of our brains that is best suited to multitasking – the subconscious.

But 'not thinking' is not the same as 'not being aware'.

If we can stay aware, staying in communication with ourselves as well as communication with the patient and all the 'other' demands on us in practice, we can stay mindfully aware of what our patient is telling us, in a balanced and questioning equipoise, ready to converge or diverge, avoiding complacency, avoiding paralysis, avoiding blindness, and creating something truly impressive.

In caring
We flow
Incepting
We start
Inducting
We think
Intuiting
We leap
Incanting
We speak
In truthing
We care

– JA

Activity 9.2: Practice makes perfect(ish) (30 minutes)

Just like every other task, becoming aware of your thinking, staying aware of it, and using it effectively in practice, becomes easier and more automated as you do it.

Try it out, firstly in theory, then actually in practice.

Go through each stage of your practice and consider what type of thinking may be most appropriate for each: convergent, divergent, or both. (Depending on what type of practitioner you are, you will have different stages at different times. For example, a family practitioner like me would typically go through some or even all of the following stages, although not in any definite order.)

- Preparing
- Clearing
- Connecting
- Sieving and safety netting
- Examining
- Formulating
- Negotiating
- Planning
- Acting
- Closing.

Now think about how you will enable your thinking to ebb and flow in actual practice, matching the correct type of thinking to the challenge of the particular moment.

How will you signal to yourself, and to your patients, that it is time to switch?

Chapter 10
Mapping

Activity 10.1: Draw health (15 minutes)

This is going to sound a bit odd, but stick with me. Right at the beginning of this book, you were asked to draw health.

Now is the time to take out the paper (if you still have it) or draw it again (if you don't).

It can be anything you like, any colour you like or any shape you like, and contain any entities you like. Just make sure it expresses broadly what health means to you.

Now draw a bicycle.

Keep hold of your creations. We'll come back to them later in the chapter.

WHAT MAPS OF HEALTH ARE THERE?

As we saw in workbook 3, no one really knows what health is. We all think we know, but when we compare notes we find everybody's definition and description is a bit different. In other words, we all have different explanatory health maps.

The maps we use to explain health and health practice are just that: maps. We all know that a map is not the real territory that it represents. Maps are neat, contained and two-dimensional. Reality is messy, sprawling and often chaotic.[37] Any metaphors that we use (including the metaphor of the infinite ring and the journey) are partial and subjective.

Unfortunately, and this is a key point, when it comes to consciousness, *maps are all we have*. Because we are conscious beings, we do not experience the matter, energy and forces of the universe simply as matter, energy and forces. Instead we

Fractals are shapes that are made up of smaller shapes that look similar to the whole shape, which themselves are made up of yet smaller shapes that look similar to the whole shape, ad infinitum. They are common in nature (for example, the broccoli florette above or the island coastlines referred to previously). Fractals were only discovered in nature towards the end of the last century, but Jackson Pollock's paintings (e.g. 'Blue Poles: Number 11', 1952[39]) have been discovered to contain fractals. So, without being conscious of it, the map he created through his art (and through his consciousness) reflected the territory that he could not have been consciously aware of . . .

take our perceptions of these simple entities and convert them in our conscious and subconscious minds into complex representations, or maps. We call these maps 'experiences' and we live inside them. From only three basic entities (forces, energy, matter) we generate an infinite number of experiences such as, for example, 'objects', 'people', 'colours', 'sounds', 'drives' and 'attractions'. This is the downside of consciousness. Consciousness ***is*** a map, and we can't see outside it.[38]

DON'T FALL INTO THE HEALTH BELIEFS TRAP

In health practice, in trying to create better health, that means we will be lost before we start if we cannot find and agree a shared map of what health looks like with our patients.

This is a hard idea to grasp. Many of us have been trained into thinking we have a very clear idea of what health is, and what our practice ought to achieve. A lot of us even have targets and tick boxes to prove when we have got there. But that's all a bit of a comfy, shared illusion.

To illustrate, go back to your creation of 'health'. If you are honest, do you think anyone else's will look like yours? Even the creation of your closest colleague, who has been trained like you, works with you, and is chasing the same goals as you will almost certainly look different.

If you are still not persuaded by that, look at your bicycle, and compare that to what your colleagues have drawn. To show I'm willing, here's my creation below.

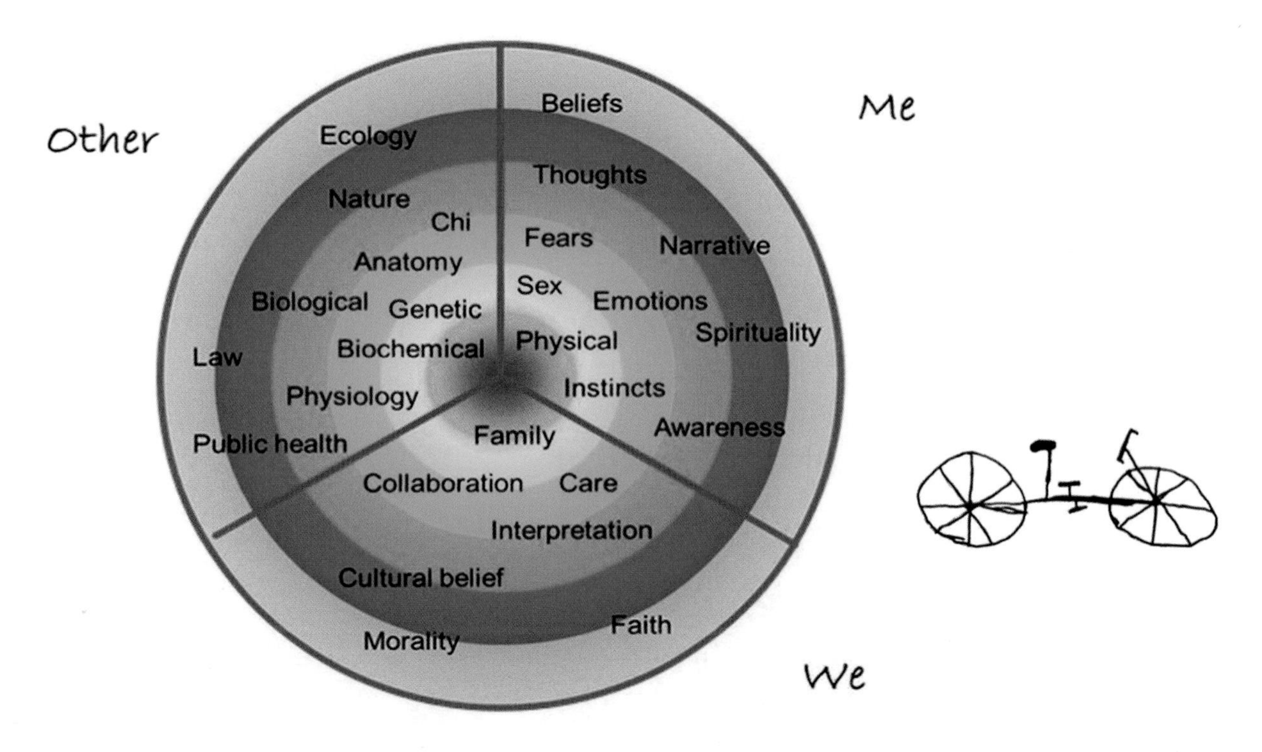

My map of health

I have chosen a circular map (clearly I too am addicted to geometric shapes) and, rather more worryingly, the colours of the rainbow to fill it in. And, perhaps because I cannot escape a deep-routed urge to categorise and classify, I have even added some concentric and radial lines to show the 'me', 'we' and 'other' entities that affect health.

The really important thing to note is that your own map will almost certainly look nothing like mine. You might have chosen a square, or a completely random shape. You might have gone 3D. It might be massive or tiny. Our maps are products of our individual perspectives and my interpretations, so they are infinitely various.

I've added my bicycle too.

CREATING AND MERGING DIFFERENT MAPS

Now I have a problem. I can't possibly imagine what your health map looks like. That's fine, because that's just like health practice. I can't possibly imagine what my patient's health map looks like either.

But let's say, for the sake of the argument, that my patient somewhat reluctantly agrees to use my basic map as a starting point, but on condition we adapt it a bit. He suggests that we both draw on it a line capturing what we each think are the most important 'bits' to health.

I'd like to pretend that everything in my map is equally important to me. As a human being, they are. But, in practice, I am a family doctor, Western trained and raised, so there are some things in my practice that I tend to put more emphasis on than others. So my line looks something like this . . .[40]

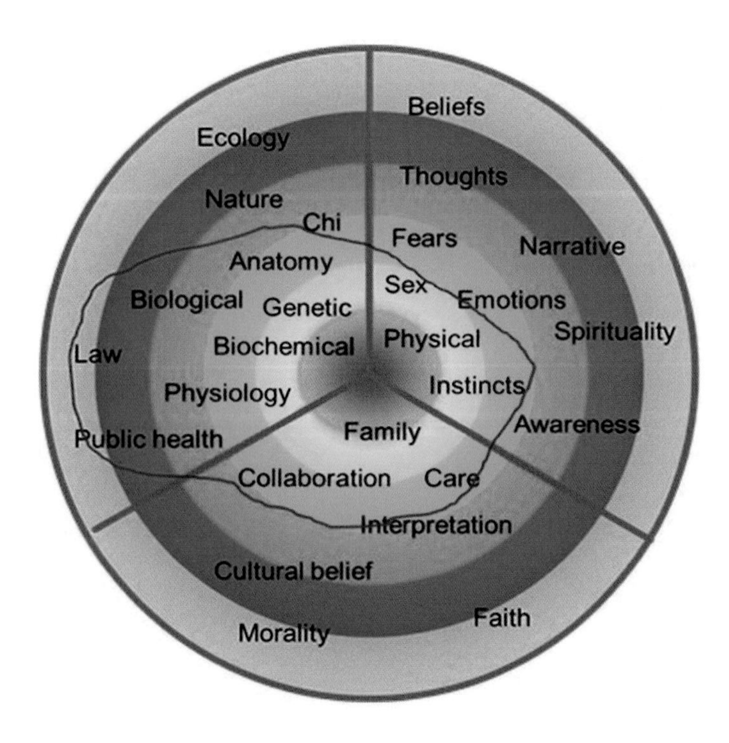

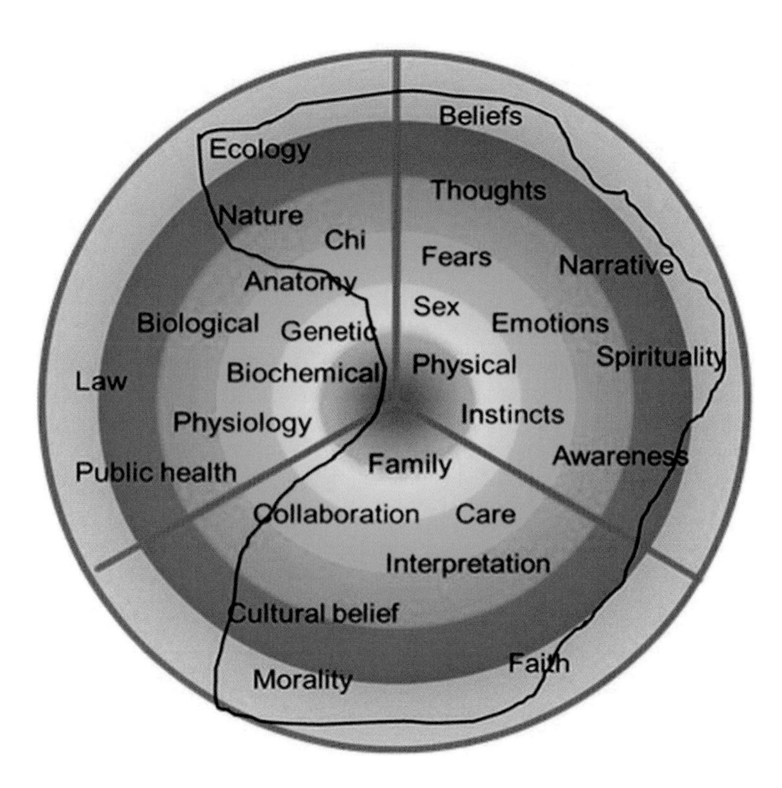

And again let's say, for the sake of argument, that my patient's line looks a bit like this . . .

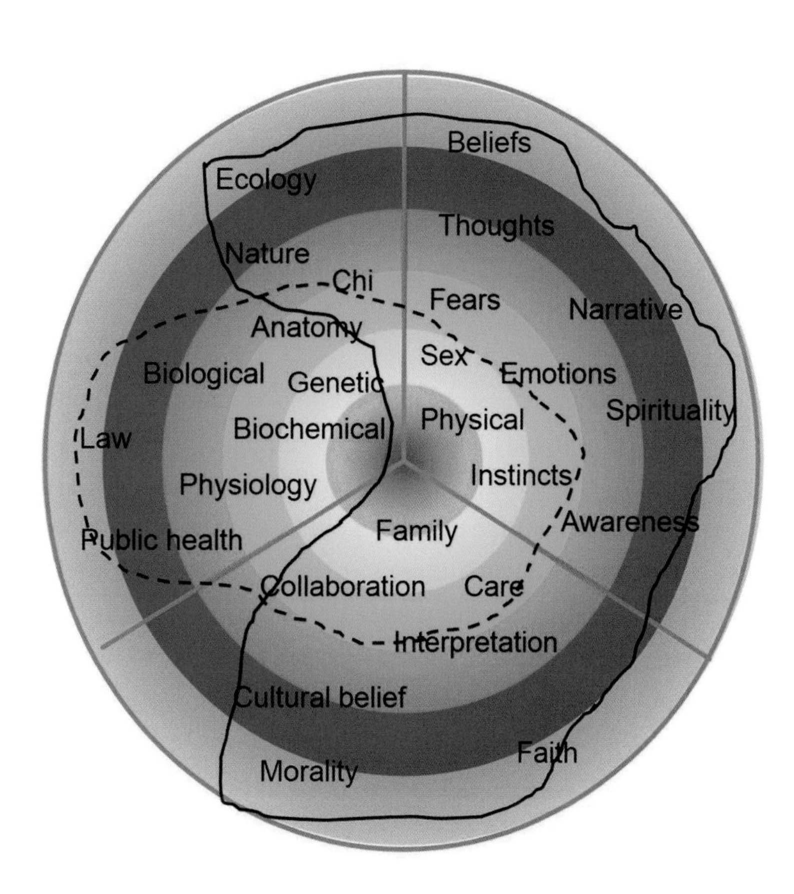

If we put our maps together, we can find some areas of consonance, and some areas of dissonance, like so . . .

CONSONANCE AND DISSONANCE IN PRACTICE

We all like a bit of consonance, because it is tuneful and restful. But sometimes we need a bit of dissonance too, because dissonance wakes us up, shakes our foundations and makes us look at things with new eyes.

While consulting, consonance and dissonance of our patient's health beliefs and our own health beliefs can be helpful or unhelpful, depending on how we look at it. For example . . .

	Helpful	Unhelpful
Consonant maps	The consultation is likely to run smoothly and efficiently, as both practitioner and patient are likely to have similar interpretations, aims and objectives.	Both patient and practitioner share the same blind spots, so they may miss potentially helpful perspectives and approaches.
Dissonant maps	Patients and practitioners, if they approach the consultation with open curiosity, may learn about new perspectives and new ways of existence that are healthier.	If the patient or practitioner is not open (they may be frightened, tired, rushed etc.), or if communication is blocked, they may run into misunderstanding, conflict and frustration.

It is difficult to overstate the importance of consonance and dissonance for effective, speedy and efficient practice. The reason that they are so useful is because most of the work we do in practice is done by our subconscious, which as we have seen operates much faster and more effectively than the conscious, analytical mind when it comes to becoming aware of and connecting with each other.

So how does it work in practice?

Well, where there is consonance, we tend not to notice anything. We just find ourselves sailing smoothly through, without even realising we are sailing.

However, dissonance is different. Almost immediately we meet dissonance, even though we may not be able to put our finger on why, we get an unpleasant sensation that tells us that things aren't right. We may not even be aware that it is cognitive 'dissonance' we are experiencing. Indeed, often dissonance emerges surreptitiously into our consciousness as negative feelings or emotions, such as anxiety, frustration or even anger.[41] We may even direct these negative emotions at the patient directly, or conversely project them onto the patient (*see* workbook 2 for more on projection, transference and counter-transference) even though he or she may be just as confused as we are about their origins.

Whichever way it is, the more mindful and self-aware we are, the quicker we will recognise that the emergence of negative emotion in practice is often a sign of subconscious dissonance. We may be tempted to plough on, and ignore it, but that is neither effective nor efficient. If we trample over dissonance, it will bite us in the behind, because (to mix metaphors) we will head off down blind alleys and circular loops, crashing into frustration, confusion, non-cooperation and dissatisfaction along the way.

On the other hand, if we sense it, and spend a minute or two exploring it, we give ourselves a much better chance of recognising it for what it is: an extremely useful tool. Why is it extremely useful? Because it short cuts us to those areas of the shared map that we need to direct our attention and effort towards. The quicker we can develop and agree shared maps with our patients, the quicker we can get on with the journey itself. In this way, dissonance can be a very powerful force in helping us to work faster and more effectively.

In practice, discovering the patient's map is incredibly simple. As long as we are mindfully aware, we can be fairly confident that we will pick up any dissonance that arises, because we will feel it as a sense of anxiety, frustration, irritation, or confusion. If we don't feel it, we can pretty much trust that it's not there, and carry on.

On the other hand, discovering the patient's map can be very difficult. If we go into the consultation already feeling frustrated, irritated or confused, we won't easily recognise the emergence of new frustration, irritation or confusion. That's one of many reasons why 'clearing' before each consultation is so important.

We just need to remember that much of our patient's health map may be subconscious, just as much of ours may be subconscious too. That means we need to be mindfully aware of our own map, skilful enough to bring our own maps into consciousness, and courageous enough to set our own maps aside for a while so that we can help the patient bring his or hers to consciousness too.

The key thing with dissonance is to notice it and to name it before it creates conflict. Once we get sucked into conflict roles, it's incredibly difficult to pull back (*see* Chapter 9 'Acting' in workbook 2 for more on this). It is therefore helpful to have a few stock phrases that we can quickly pull out in the split second between the emergence of dissonance and the development of conflict. Some examples might be as follows.

- Different people look at health in different ways. Do you think it might be useful to explore our views briefly?
- I am sensing we are not quite understanding each other, which is quite common, so don't worry. Shall we take a little time trying to agree what the way forward might be?
- I don't know if you've noticed, but I think we might be becoming a little frustrated with each other. That is quite understandable, and usually just means that we don't fully understand each other. Do you think it might be worth exploring that a little?
- You used an interesting word (or image) there. I wonder if you could say more about how you see this whole problem, and what might be causing it?

Once we have picked up the points of difference, we can choose either to set them aside as understandable and allowable differences, or (if we feel it is going to be difficult to continue unless they are addressed) we can choose to bring them out into the open and explore them.

If we do that, one of two things will happen. Either we will reach a degree of consonance that we are both happy with (we don't have to agree on everything, just the things that are going to matter to our co-creation). Or we will not. If we can't reach consonance we either have to accept our co-creation is not likely to be very effective; or we can agree that the patient may be better seeking another practitioner's help.

In an infinite world, not all paths converge. Divergence is at the heart of creation and evolution.

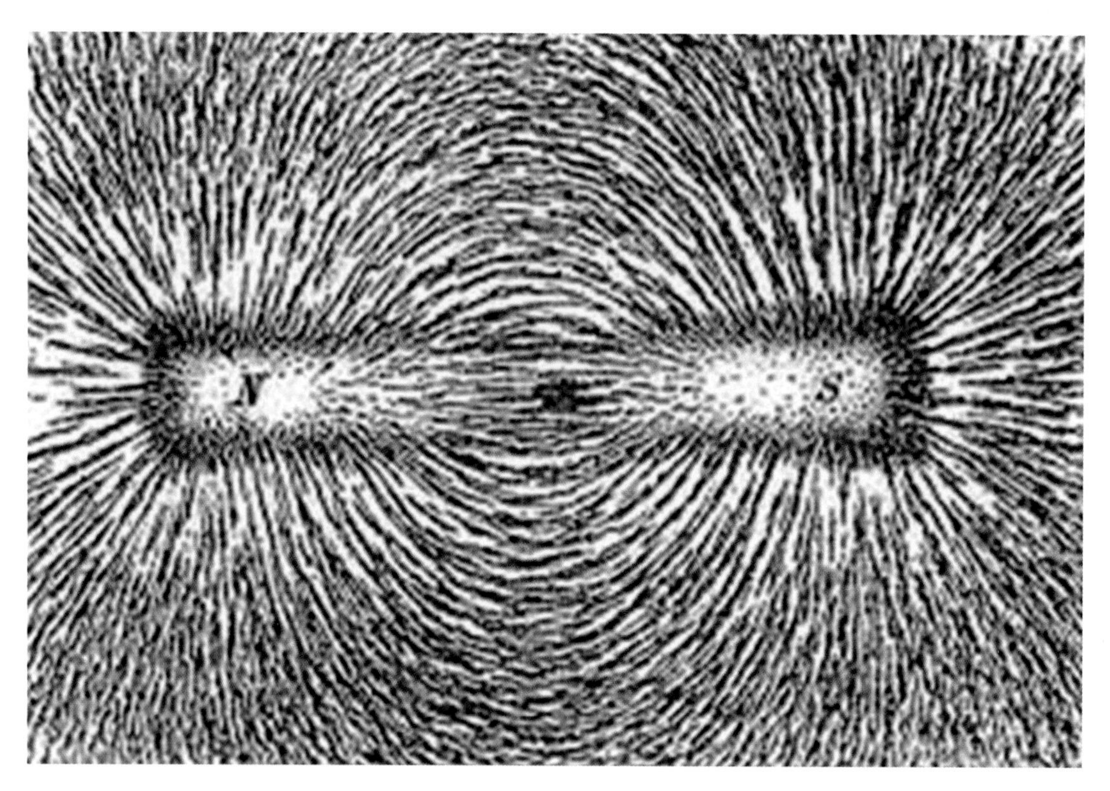

Magnetic field – poles of a magnet co-create fields within which an almost infinite number of paths are possible, some convergent, most divergent.

If our paths diverge, that is not a problem. We are not alone. We are all team players, so we can signpost or refer our patients on to others who we think might be able to help where we cannot. We don't need to be too protective or think that somehow we 'own' our patients.

There are more than enough to go round.

INTEGRATING MAPS INTO PRACTICE

As we can see, each of us will hold perspectives of existence and of health that are more or less relevant for other people. In the same way each of us will have ways in which we can teach the other. Therefore health practice can be a process of mutual learning and creation, both for patient and practitioner. But, if we clash and cannot achieve consensus, health practice can be a process of mutual frustration and recrimination.

If we can find and agree a shared map, our creation of better health will be more satisfying, more effective, and considerably quicker. If we cannot, the process will be circuitous, unhelpful, frustrating, slow and ineffective.

Constructive or destructive, the outcome is largely in our hands. If we can be open, and enable our patients to be open, we have the best chance of reading each other and creating a map we can share. If we are closed, or we cause our patients to close up, we will not.

Through communication comes understanding. It is not that life is like this, or like that. I might be wrong, you might be wrong. We all might be wrong. If we are honest, no one really knows what health is.

It is more that we have the choice to accept the freedom, and the responsibility, to create our health, indeed our very existence, in a way that makes sense for both of us.

O how may I ever express that secret word?

O how may I ever express that secret word?
O how can I say He is not like this, and He is like that?
If I say that He is within me, the universe is ashamed:
If I say that He is without me, it is falsehood.
He makes the inner and the outer worlds to be indivisibly one;
The conscious and the unconscious, both are His footstools.
He is neither manifest nor hidden, He is neither revealed nor unrevealed:
There are no words to tell that which He is

– Kabir

Activity 10.2: Welcoming dissonance in practice (a few hours)

Dissonance makes us uncomfortable, but we can choose to see it in one of two ways: as an irritating distraction or as a welcome friend who has come to teach us something.

For the next few sessions you are in practice, try to keep an eye out for the emergence of dissonance within you. When you feel it arise, metaphorically smile at it and welcome it with open arms. Consciously set your own agenda or concerns aside for a while, just for a minute or two, and use some of the questions above, or your own, to explore the patient's model.

Then, if you feel up to it, try briefly sharing your model with the patient and see if you can negotiate and co-create a shared model.

If you can't, don't worry; it's not possible to agree on everything with everybody. If you can, congratulations; you have both learned something and you are likely to co-create much better health.

Chapter 11
Negotiating

Activity 11.1: Your ideas, concerns and expectations (20 minutes)

Imagine you have a headache. It's pretty severe and you have had it nearly a month. You decide to go and get help from a practitioner.

- **What ideas cross your mind about what it might be?**
- **What concerns do you have over what might happen if you don't do something?**
- **What expectations would you have of the practitioner you choose to see?**

Now imagine you are someone completely different: different sex, different age, different ethnic background, and different culture. Repeat the process, and compare any differences in your two sets of responses.

CO-CREATION CAN ONLY BE DONE IN PARTNERSHIP

Generally speaking, most of us have reasons for travelling. We may need to be somewhere else, or we may be unhappy where we are, or we may just enjoy travelling along. As our practice is a shared journey with our patients, it may be useful for us to have a good idea of each other's motivations.

I might wish to go here. You might wish to go there. We have different 'agendas'.[42] Failing to work out each other's agendas can lead us into taking longer or more tortuous paths than we need to, or even cause us to disappear down dead-ends.

On a shared journey it always helps if we pull together (preferably in the same direction)[43]

NEGOTIATING AGENDAS

You may be familiar with this kind of situation, which all too often happens to me . . .

Me: Well, Mr Smith, you came to see me with a sore back and chest pain 6 months ago. I was concerned about your chest pain so we've been through your history and examined you thoroughly. We've run blood tests, exercise ECG, angiography, gastroscopy and chest X-ray, all of which have been normal. We've discussed coping mechanisms, arranged for you to have bereavement counselling and some CBT in case there was a psychological cause at play. We have referred you to the exercise inclusion scheme and advised you on your diet. And I'm afraid, after all that, there is really nothing more we have to offer.

Mr Smith: That's OK doc. I've had the chest pain for years and it's never really worried me. I just take some peppermints if it plays up. All I really wanted was a sick note for my sore back.

When patient and practitioner think about the journey ahead, they tend to ask themselves some questions,[44] for example:

- **ideas** about where we are
- **concerns** about what might happen if we stay where we are
- **expectations** of what might happen if we move somewhere else.

Given that these three things (which nicely abbreviate to 'ICE') are quite important in influencing where we want to end up, we would think that these are a very good place to start our discussions. Apparently, we would be right. As we saw in workbook 2, there is a lot of evidence about the importance of getting the patient's ideas, concerns and expectations out early, and revealing our own ideas, concerns and expectations early too.

We might also think it would be easy for patients (and practitioners) to get their agendas out into the open nice and early, but here we would be wrong. Apparently, it is not easy at all.

Both patients and practitioners[45] may be reluctant to share their agenda, for various reasons. Perhaps we don't wish to be disrespectful, or frightening, or pushy; or perhaps we are afraid of 'naming' our fears; or perhaps we do not fully understand our own reasons for being there; or perhaps we are not aware of our own subconscious drives and motivations.

Given this apparent difficulty with even recognising, let alone sharing, our individual agendas, the more mindful and self-aware patients and practitioners could be, the better. Some questions that may help our patients (and ourselves) to become more mindful and self-aware would include the following[46]

Ideas:

- 'Could you tell me about what you think is causing it?'
- 'What do you think might be happening?'
- 'Do you have any clues or any theories as to what is going on?'
- What do you think might lie behind or be causing your problems?

Concerns:
- 'What are you concerned that it might be?'
- 'What's at the back of your mind?'
- 'Is there anything particular or specific that you were concerned about?'
- 'What was the worst thing you were thinking it might be?'
- 'In your darkest moments, what do you think might be going on?'

Expectations:
- 'What were you hoping we might be able to do for this?'
- 'What do you think might be the best plan of action?'
- 'How might I best help you with this?'
- 'You've obviously given this some thought; what were you thinking would be the best way of tackling this?'

NEGOTIATING ROUTES, WAYPOINTS AND END-POINTS

In an infinite ring, any start-point, waypoint and end-point is possible. One person's end-point may be another person's way point, or even their start-point.

We don't want to arrive at what we think is our destination only to find that the patient wanted to go somewhere else entirely.

So, in any one consultation it is difficult to balance what to include and what to leave out as, for sure, we cannot include everything. It is even harder to deal with blind spots as, by definition, we don't know what we don't know.

For example, if a man presents to us with chronic headaches, we may take any one or more of several possible routes: molecular routes (use of painkilling drugs), naturopathic routes (eat and drink in a more balanced way), physical routes (improve posture, relax neck and shoulder muscles), behavioural routes (cut down on alcohol), psychological routes (express and explore causes of muscular tension and expression/experience of pain), environmental routes (use protective gear to stop inhaling fumes), or spiritual routes (what is pain anyway, and isn't some pain purposeful and meaningful?).

It is not that we wish to generate a strict contract up front because of course things can change during the journey. Furthermore, subject to what crops up along the way, we may have to renegotiate our end-points and routes as we go along. Nevertheless, it is useful to pause, discuss and 'signpost'[47] our agreed route both at the start of the consultation and at various waypoints along it. Some useful phrases that might help us agree routes, waypoints and end-points might include the following.

- 'So, where would you like to start?'
- 'Would it be helpful if we looked at this a different way?'
- 'Well, I wonder where all that leads us?'
- 'This reminds me of the time . . .'
- 'I can see that, but perhaps another perspective might be . . .'
- 'Perhaps you could recap what we've covered?'

A WORKED EXAMPLE

As an example, let's say a 30-something, white, British, male patient presents to me with a 'headache'. It becomes rapidly clear that he wants his headache to go away and he thinks I can help. We are both of a similar cultural group, and so it is likely, though not definite, that a biomedical 'health map' is OK for him.

His open agenda: to get rid of his headache.

My open agenda: to help him feel better within the resources I have available.

His hidden agenda: what he doesn't tell me is that his uncle has just been diagnosed with high blood pressure. His mum also died last year of a stroke, and he thinks that may have been caused by high blood pressure.

My hidden agenda: what I don't tell him is that I would want to rule out any sinister 'medical' causes of pain, or that he looks like he might enjoy using recreational drugs. I also don't tell him I need to run a 'well person screen' on him as we haven't yet hit our target points (and points means prizes); that I really need to get away on time as I have a meeting with fellow trainers, and that my blood sugar and caffeine levels are critically low. Finally, I don't tell him my hope that, if I am quick, he will probably be happy to let me do his health check and we'll both finish in time for me to catch a coffee before the next patient.

(I think I might leave my thoughts about his possible recreational drug use until next time.)

The quickest and easiest route to these end-points is probably to check quickly for any physiological or anatomical causes for his headache and then to prescribe a pill. It's not great, but it probably won't do harm (given the evidence that we are aware of), it will get to the objective (stopping the headache) and it will be very quick and cheap. It's good enough.

On the other hand, neither of us is going to feel that good about him going away with only one of his objectives met. He is going to be back soon as the other issues start to express themselves into his existence. I'm going to get bored stiff if I practise like that for any length of time, and my preferred (if often deluded) image of myself as a caring and conscientious doctor is going to need considerable retouching.

But, let's say I have more time with him, and my caffeine and sugar levels are adequate. It could be fun to go on a bit of a journey together. We both might learn something, we both might discover and play with some different perspectives and ideas, and we might create something that we can both build on. For example . . .

Me: So, where shall we start? (Signposting and handing over control.)

Him: Well (pauses, looks down – thinking, maybe censoring) . . . I've had this headache (pauses again, looks down).

Me: (Smiles gently, nods encouragingly, says nothing.)

Him (after a few seconds): Well, it's been troubling me a lot.

Me: 'Troubling?' (Interesting choice of word, many possible meanings. What's his narrative? Try simple repetition to demonstrate listening. Try inflexion to invite him to say more.)

Him: Yes, I've been getting it most days, and it pulls really badly behind my eyes (gestures with clenched fist in front of forehead – anger? Pain? Repressing something?). It usually settles after I take some painkillers, but then it's back the next day. It's beginning to interfere with my work (suddenly bangs forehead with clenched fist –cue) and it's bothering me at home too.

Me: Hmm. (There are a couple of threads there, plus more interesting words – 'bothering' and 'interfering'.)

Him: Yes, so that's really why I've come (tails off, signposting to me he has finished for now).

Me (thinking – seems to be no drama here, it's pretty much adult-to-adult stuff, so let's keep it that way): OK, so it seems to me that you are getting troublesome headaches which are bothering you and interfering with your life (repeating, use of same metaphor, and summarising). Was there anything else you came about today (digging for hidden agendas, desire not to head off down blind alley, but fairly direct, addressing his cognitive thinking)?

Him: Well, I haven't had my blood pressure checked for a while.

Me: Yes, that's a good idea (affirmation, encouragement to contribute). I wonder if you might be thinking blood pressure and headaches may be connected (attempt to explore understanding and look for hidden agenda, but fairly direct, using 'thinking' rather than say 'wondering' or 'worrying').

Him (loosening up, happy to be complimented and pleased that the doctor has picked up the link): Yes, my uncle had blood pressure and (pauses, dissonance, looks up and away) . . .

Me: (Says nothing. Smiles. Leans slightly forward. Eye contact from lower position.)

Him (after a few moments, looks back at me, takes deep breath and opening gesture with his hands): Well, I'm probably being daft . . .

Me (smiling): I doubt that. (Change position suddenly to create dissonance.) I'm sure you would only mention things that are important to you (hidden command).

Him (looking away again, but now more still, and looks sad – internal processing, experiencing but hiding emotion, maybe remembering).

Me: I wonder if it would be helpful if I go through what I am thinking (signposting – it's OK, you're still in charge and we won't go anywhere you don't want).

Him: Erm . . . (thinking, not sure, trying to assess threat and benefit, deciding) OK.

Me: Often, people come to me with headaches as they are worried it might be due to something . . . (deletion, generalisation, pause).

Him: Well, perhaps . . . (I'm sort of OK with this but I'm not sure).

Me: Headaches can have many causes, mostly not sinister. Sometimes they can be caused by physical things, but the body can be quite clever at telling us what is going on inside (teaching, trying to avoid jargon). Some patients (generalise, less threatening) that come to me (subtle cue that I am experienced/confident/competent in what's coming next) get headaches because they are hurting inside, maybe through sadness, or grief . . . (suddenly raise eyebrows and say last word with slightly more accent to cause dissonance, it's a strong cue for him, pause to allow him to go inside and process then decide if that's a route he wants to take).

Him (in slight trance, internally processing): Well, my mum did die of a stroke last year . . .

Me: Oh, I'm very sorry to hear that. Pause. (Wait, wait, wait. Don't interrupt. Don't. Really, don't.)

Him: (Sits back, gaze down and to right, looks emptied and drained. Says nothing but blinks frequently – given up his resistance to 'trouble', 'bother' and 'interference'?)

Me (thinking – he's open, gently test how open): Would I be right in thinking that you may be worrying that might have been to do with her blood pressure? (Summarising, making links, suggesting, but direct, interrupting and appealing to cognitive level.)

Him (relieved, bit embarrassed): Yes, I suppose so.

Me: Pause . . . (time to give him a break, let me take the reins for a while). Look (hidden command – come out and pay attention) why don't I summarise where I think we are at (signpost and hidden command). (Brief look down and pause – change sense, change pace, generate dissonance, possible micro trance.) I understand that you are worried about your blood pressure, particularly because you lost your mum to a stroke (I hear you). So perhaps we should make sure we check that. From my perspective as your doctor, it would probably be worth me examining you and possibly running some tests, depending on what I find when I examine you (don't worry, I will be thorough). I'm not expecting to find anything (reassuring). It won't take long and I'll maybe do a quick health screen if that's oaky? (Signposting, negotiating, but almost commanding.) Perhaps it might be worth us looking at what you do to relax and have fun (offer new agenda, possible opening to drug use)? But, before we move on, it may be wise if we quickly both think if there might be anything else that might be causing it that we haven't thought of yet (invitation to revisit the map for blind spots). And, perhaps most importantly, I do also wonder if you'd like to talk a little bit more about your mum (sensitive issue, left until last to emphasise permission, offer of new agenda).

Him (thinking, remembering, feeling sad): Yes, well . . . what do you think? (negotiation)

Me: I'm happy to help with any of those (you're in charge, as you've come to see me). I'm not sure we can do it all in one go, but we can probably do some now and some another time (signposting, agreeing which problems to cover and which to leave, reminding him there are limits to my resources – I also have to meet my agenda and be ready for the next patient in 5–10 minutes). But please don't think I'm only here to help with physical problems (trying to counterbalance any belief he has that doctors may not deal with psychological or social issues).

Him: OK. It would be good if you could check my blood pressure, and if you could reassure me that my brain is OK. Maybe I'll book back in to see you about the rest later on.

THE ROAD TAKEN

This exchange has helped us to achieve a lot. We've recognised and affirmed each other's agendas, we've suggested some end-points and waypoints, we've discussed the route and we've recognised that we have to operate within resource constraints.

Let's look at this in a visual form, using our map again.

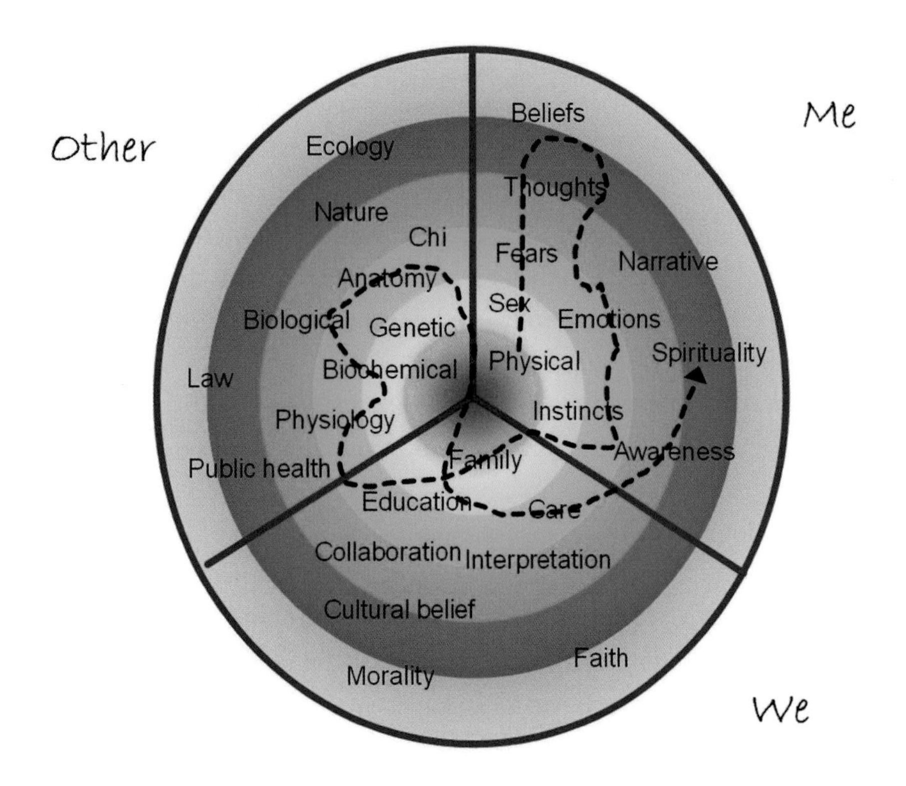

A possible agreed route through our integral map

There are a number of waypoints that we have agreed to, and some of them we have already passed. We have heard about his physical problem (pain), begun to explore his fears, thoughts and beliefs; we've generated and affirmed some emotions.

We haven't really looked into the ecological or epidemiological causes of his pain, an omission which might turn round and bite us. Also, we have kind of assumed a shared culture and belief system, although our evidence for that assumption is not strong.

But we can't do everything. We can only be good enough.

And, anyway, we've achieved a fair amount already. We are only 3 minutes in and we've looked at his fears, thoughts and beliefs; he has expressed and I have accepted some emotion at his situation, and we have increased our awareness of each other and of the kind of things we might wish to look at. We have begun to plot out the route from here: which will be checking out any physiological, anatomical and biochemical causes for his pain before looping back to consider the effects of his family connection, his home setting and his bereavement.

We may still have an opportunity to wander through some educational areas and (if we have time and are feeling brave enough) maybe even have a little flirt with the edges of the 'spiritual' area. That could be useful, educational and interesting for both of us.

And we can still use the painkillers if we want.

A GOOD ENOUGH JOURNEY PLAN

None of us is perfect.

When we choose our path, we may not always choose most wisely. But it seems reasonable to suggest that our journey will have a greater chance of success if we dedicate ourselves to our patient (so that we use our energy in a focused and targeted way), if we stay mindfully aware (so that we can identify agendas early), if we communicate effectively (so that we can discover agendas and negotiate routes and end-points), and if we act skilfully (so that we guide our shared journey as efficiently and effectively as possible).

To pursue the map metaphor a little longer, a more mindful practice using a more skilful range of techniques may well end up at the same place as a less mindful one (as is illustrated in the case above). But a more mindful process will take into account a wider range of perspectives and so generate a broader, richer and deeper co-creation; and therefore a broader, richer and deeper map of that co-creation. With a better map, we have a greater range of routes and therefore a greater chance at arriving at end-points that are healthier for the co-creation; and so for both the patient and for me.

This opening up is a process of opening awareness, for both patient and practitioner. A more open awareness will see more strands, hear more notes, and taste more flavours. In some ways this makes integration and balance harder. But it makes the end result richer, fuller and more satisfying for everyone: a symphony rather than a melody, a work of art rather than a sketch.

And it doesn't take long. The whole interchange above would probably take three minutes at normal pace. And it's quite enjoyable and interesting. That's pretty efficient and effective practice.

But, it's not going to be perfect.

What neither of us knew or managed to work out was that his headache was caused by working with potentially toxic gases in a paint shop where the workers don't always wear correct protective gear. I missed the cue about his work and so neglected to explore that environmental area.

Living and learning.

Road Not Taken

Two roads diverged in a yellow wood,
And sorry I could not travel both
And be one traveler, long I stood
And looked down one as far as I could
To where it bent in the undergrowth;

Then took the other, as just as fair,
And having perhaps the better claim
Because it was grassy and wanted wear,
Though as for that the passing there
Had worn them really about the same,

And both that morning equally lay
In leaves no step had trodden black.
Oh, I marked the first for another day!
Yet knowing how way leads on to way
I doubted if I should ever come back.

I shall be telling this with a sigh
Somewhere ages and ages hence:
Two roads diverged in a wood, and I,
I took the one less traveled by,
And that has made all the difference

– Robert Frost[48]

Activity 11.2: Co-creating a journey plan (1 session)

Try co-creating shared maps with your patients in practice.

For the next few times you see a patient, consciously explore the patient's ideas, concerns and expectations early on in your consultation, before you head off into your own assessment and management of the patient's 'problems'. If you feel yourself reacting to dissonant ideas, behaviours or beliefs, try not to engage with your feelings but mindfully allow them to melt away. Keep going until you feel confident that you have a fairly clear impression of the following.

- Why they have come: Why here? Why now? Why me?
- What do they think lies behind their problems, whether physically, psychologically, socially, spiritually or culturally?
- What do they fear will happen if they don't get help?
- What do they want or expect me to do about it?

Try to note the time it takes to do this, and then see if it makes you any more efficient or effective in your practice.

Chapter 12
Letting go

Activity 12.1: Letting go (30 minutes)

Take some time out to think. Take a pen and paper and write two lists:

- External censors: all those things that you feel constrain and censor you in practice. It may be your colleagues, regulations, time, resources or targets. Whatever. Just list them all down.
- Internal censors: all those internal things that stop you expressing yourself freely, expressively, and creatively. It may be your sense of embarrassment, duty, perfectionism, beliefs, traditions, culture. Whatever. Just list them all down too.

Look at your lists. Then think to yourself: what I am here for? If we can't even express ourselves in our practice, our career, when can we express ourselves?

Now put the lists to one side, or throw them away, or even burn them.

Allow your body to relax and your mind to clear. For a few minutes visualise yourself as completely free in your practice. Free from constraints and censors, practising how you would like, how you think it ought to be done. Creating and expressing.

What's stopping you?

This is it.

Nothing happens next.

Before we can go about creating anything (better health included), we need to create some space in which to create – our canvas, if you like. The clearer and cleaner our canvas, the greater the potential for creativity. This is the same for creating 'better health' as much as for 'better art' or 'better poetry'.

The problem is that open space and time are so rare in practice. We like to blame others for this, and there is no question that there are lots of tyrants out there, but perhaps one of the reasons we are so busy is that we secretly prefer that to the opposite. Maybe emptiness empties us, and we prefer to be full.

Perhaps sometimes we react to this fear of open space by allowing 'trying' to become part of us. As health practitioners we often try so hard that eventually we become trying, to ourselves and to others. But the reality is, despite all this trying, we will all fail in the end, because everything comes to an end, and everything empties out.

Nothing happens next.

THE ABSOLUTE IMPORTANCE OF LETTING GO

It is so very hard for us to let go.

It seems somehow unfair. We have taken all this time, dedicated all this effort, acquired all this knowledge and experience; played (mainly) by the rules, hit a fair number of targets, ticked most of the boxes, maybe even developed a good reputation among our colleagues and patients along the way.

Perhaps, if we try harder, we can keep it going, maybe even get better?

The problem is, in the moment, none of that matters. It is all in the past or future. It is all outside the space and time we are currently in.

As self-conscious beings, we are our own canvas. When we create, when we sit with our patient and co-create, the moment is empty – until we fill it. If we allow our canvas to become messy and our space to become noisy, the next moments that we create will be poorly formed, half-baked and stunted.

So letting go is not giving up. It's not giving anything up. It is just humbly recognising and surrendering to the extraordinary power that we already have. It is an incredibly difficult concept to get our heads round, but the fact is that we are able to self-create and co-create, not metaphorically, but quite literally.

On top of this quite universal and astonishing human ability, we practitioners have added more powers: for example, experience, knowledge, skills, dedication and tenacity. These are powerful tools which can enhance and inform our creativity. But they are just that, tools. They are not the thing, the creation, itself.

We are expressions of the universe, and so the universe will express itself through us. When it comes to the moment of creation, we just have to let go, and let it happen. By letting go we do not let go of our power, we just let go of constraint, of censorship, of fear: fears of what we might not achieve, or of what others might think, or of how it might end.

Our doubts are traitors, and make us lose the good we oft might win, by fearing to attempt.

– *Measure for Measure* (Shakespeare)

Sounds like a load of baloney to me.

I agree. But read on.

Just in case it isn't.

(RE)CREATING OURSELVES

We have seen how practice can exhaust us, burn us out. We have discussed how we may be sometimes tempted to overcome this problem by seeking to separate ourselves from work, and achieve what has become known as 'work–life balance'.

But, can our 'work' really be somehow separate from our 'life'? Even a moment's reflection throws this into doubt. We live through our practice, and our practice lives through us. If we try to amputate it, surely we will just end up amputating a core part of ourselves?

From another perspective we may be able to stay whole. Our practice is a fundamental part of our existence, and probably one of the parts where we feel we can be, at times, truly expert. Our practice *is* our life.

If we think of it like this, we can start to see our practice as a co-creation which is not just for the benefit of our patients. It is also a blessing for us. It is a co-creation through which we can learn, develop and express ourselves, just as a musician composes or an artist paints.

Our practice is our canvas.

We can start to create that practice by allowing it to create us. That is, we can allow it to act as a tool and a journey for our own learning and development, helping us to express ourselves and the universe in ever more colourful, harmonious and beautiful ways.

OK, that all sounds very nice. But when was the last time you cleared up a puddle of vomit?

A BRIEF MEDITATION ON DEATH

Despite appearances to the contrary, the above is not such a facile question.

As it happens, meditating on and immersing ourselves in illness, physical corruption, decay and death are very important activities, not just for us as practitioners, but for us as human beings.

Many contemplative traditions encourage us to face up to and accept the fact of our own steady decay and corruption into eventual death. This is not just from morbid curiosity, but because it helps us keep perspective and value what is important in our lives.

It sometimes is helpful to meditate on endings and death, and so we will take the opportunity to try that now.

Please feel free to skip this section if you are feeling too raw, or if death is too recent.

Relax your body as normal, and watch as your breathing eases and your mind settles.

When you are feeling peaceful, imagine yourself driving at 100 miles per hour along a motorway, with the roof down, your hair in the wind and a blue sunny sky above. Then imagine that, suddenly, out of nowhere, a concrete wall appears irrevocably and unavoidably in front of you, and watch as you crash, fatally, into it.

Watch as your body reacts. Maybe your heart rate and breathing increase, your muscles may tense up, and strong emotions may appear. Don't engage but keep watching and waiting, and they will gradually clear and settle.

Now remind yourself that this is how life is. We are all going to hit that wall, one day. We just don't know when.

Watch as your body reacts again: your heart, your breathing, your muscles, your emotions. Keep watching and waiting, and they will gradually clear and settle.

Now allow your attention to start to rest on your own physical and emotional infirmities. Become aware of which parts of yourself are already corrupting and decaying.

Watch as you react again and wait. As before, everything will settle.

Try to visualise yourself before you were born. How did it feel? Was it frightening not being alive? Or did it feel simply peaceful, empty and calm? Now imagine yourself after you have died. How does it feel? Is it frightening not being alive? Or does it feel simply peaceful, empty and calm?

Now smile at yourself and become aware of the life that is in you, that is you. Feel the air in your lungs, the fluidity of your body, the strength in your bones, and the fire in your heart. Congratulate yourself on the miracle of being alive.

And dedicate yourself to using that gift wisely and compassionately in your practice.

Life and death[49]

HEALTH PRACTICE AS SELF-PRACTICE

Our practice can be many, many things.

> 'Practise' v.tr.
> - To do or perform habitually or customarily; make a habit of
> - To do or perform (something) repeatedly in order to acquire or polish a skill
> - To work at, especially as a profession
> - To carry out in action; observe.
>
> Practice n.
> - A session of preparation or performance undertaken to acquire or polish a skill
> - The skill so learned or perfected
> - The condition of being skilled through repeated exercise
> - The act or process of doing something; performance or action
> - Exercise of an occupation or profession.
>
> Archaic:
> - The act of tricking or scheming.

We can trick ourselves, and scheme with others, for so long, but not forever. We are who we are and the truth will out.

Our practice is many things, but it is always one thing. It is always part of our lives. We cannot separate our work from our life. Our practice is always a working out of our beliefs, our values, our training, and our wishes. In that sense my practice ***is*** my life.

Because my life may be many things, all at the same time, I may see my practice as many things, for example:

- my art
- my music
- my science
- my exploration
- my analysis
- my yoga
- my worship
- my poetry
- my dedication
- my everything
- my nothing.

They say our greatest fear is not of what we may not achieve but rather of what we may achieve. We cannot each be a genius. We cannot create perfection. We will all make mistakes. Many mistakes.

But as practitioners, in this moment, we can create something very special out of the emptiness of the moment, and that thing is 'better health'. Most people value their health almost as much as they value their loved ones, so what we create is both highly valued and extremely valuable.

We can choose to practise health, or we can choose not to. But we cannot stop creating, because we are conscious, and to be conscious is to create.

So what we can say about our practice is that it touches either everything or nothing. It derives from every part of our existence (and touches on every part of our existence) or it does not exist at all.

Either way, that makes health practice either an impossible challenge or a waste of time. If you are reading this book you have probably, like me, decided to go for the former.

We are all alive, but we have to continue to create ourselves, every moment. If we don't fully create ourselves, we don't fully live. It's a heavy responsibility. In fact it's such a heavy responsibility that we often pretend we don't have it. The fear of our own potential is too great, so we hide behind a self-built prison, pretending to ourselves we are the victims of fate and fortune.

But this is it. This practice is our life. It creates us and it creates our health, just as it creates our patients' health. Our health practice is our life practice, and our life practice is our health practice.

We are the biggest anchors on our own creativity. If we can let go, just for a moment, we can truly live, truly create.

To Be Alive

To be alive—is Power—
Existence—in itself—
Without a further function—
Omnipotence—Enough—

To be alive—and Will!
'Tis able as a God—
The Maker—of Ourselves—be what—
Such being Finitude!

– Emily Dickinson

Activity 12.2: Life practice (30 minutes)

Go deeply into peaceful meditation and visualise yourself, in practice, in life, as you really are.

Just as when we look in the mirror, the chances are you will initially see things you don't like. If that happens, don't engage or lose heart. No one is perfect. Wait patiently for those emotions and thoughts to drift away.

Engage in some meditative behavioural therapy. Stay in meditation but smile, broadly.

While you are smiling, start to see yourself in a positive way. Start to notice those things that you do well in practice, in life. Begin to see your strengths, and even your expertise. Look back, and see how far you have come.

Now zoom out a little and look at the context, the background. Could you have become who you are without your practice, for better or for worse? Understand that, as you have created your practice, it has created you.

Stay peaceful but engage your rational mind. If you can practise expertly with patients, is there any reason you cannot practise expertly with yourself? If you can create healthy patients, is there any reason why you can't create a healthy self?

Choose one or two things that you are going to change to make your health practice more of a self-practice as well.

Smile and return to now. Be alive.

Chapter 13
Models and reality

Activity 13.1: Your practice vs model practice (30 minutes)

Consider how you have been taught to practise. In that teaching, what was held up as the 'ideal' practice? What models were used to explain how you could or should practise?

Have a think about different models of practice that you have come across. How do they interrelate? How useful are they?

Now consider how you actually practise. Does your actual experience match the models? If not, why not?

MODELS AND REALITY

We are nearly there. We are almost at the point where things come to fruition, but before we do that, there is one more step.

It's time to get practical.

Whenever we create something new, it is often helpful to create a model first. Modelling helps us to simplify, break down, analyse and practise the steps required to create our creation. It is really quite unwise to attempt to create something as complex and sophisticated as better health without practising with models first.

But there is a problem with models. Models are neat, simplified and contained, whereas real life is messy, complex and infinite.

The fact that we don't exactly know what health is may be seen as a rather embarrassing but ultimately irrelevant issue for health practitioners. If it walks like a swan, and if it quacks like a swan . . .

Because of its inherent messiness and complexity, learning how to be a health

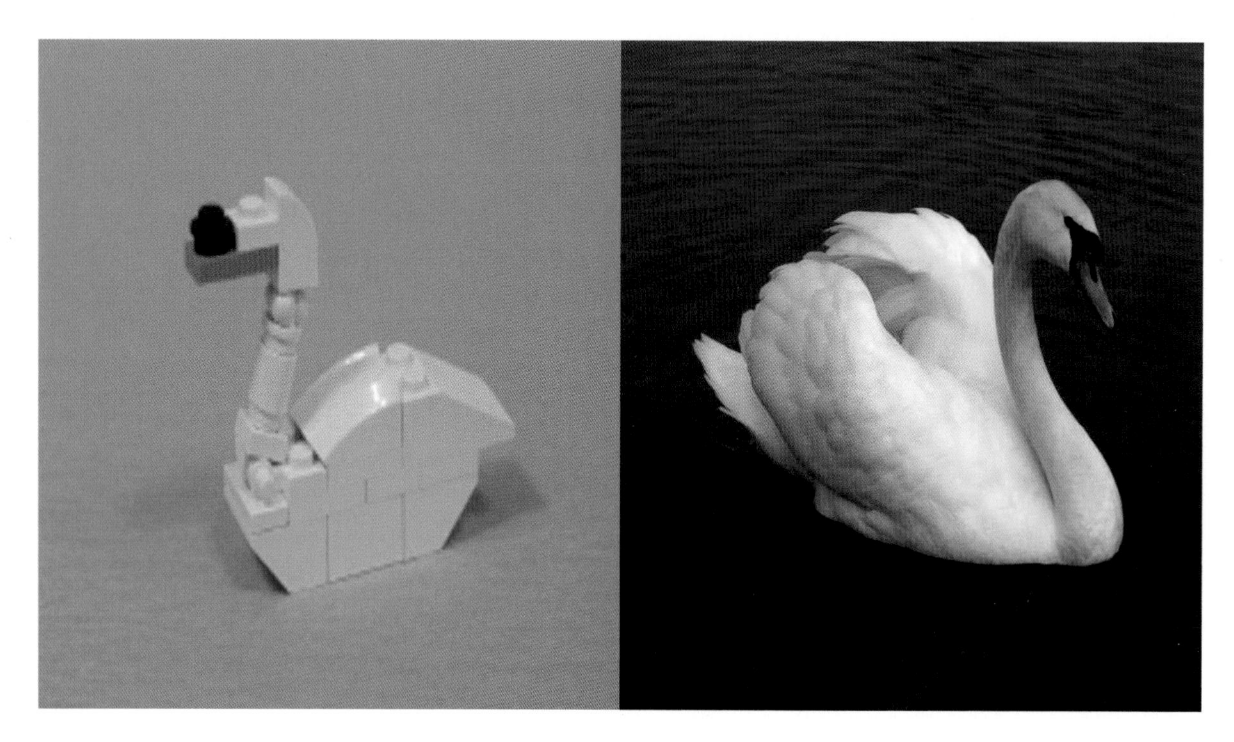

Models can be useful, interesting, and even beautiful, but they can never match the reality of existence

practitioner can be a daunting, maybe even an overwhelming task. Perhaps for that reason, over the years, various practitioners have created many different simplified and codified 'models' of practice. These models are usually intended to help us to understand better and to practise more effectively. The following are examples.

- Models of medical practice 'as' biomedical, psychotherapeutic, technical, anthropological, investigative, transactional, interventional, narrative or 'inner' entities.[50]
- Models of nursing practice 'as' comforting, guiding, assisting, attending, communicating or modelling.[51]
- Models of psychotherapy practice 'as' analysis (Freudian or Jungian), gestalt, drama, interpersonal, family, narrative, cognitive, behavioural, hypnotherapy, play therapy or neurolinguistic programming.[52]

THE MODEL AS A TOOL

All of these models can be extremely helpful, in that they give us an opportunity to practise our skills before we actually 'go live'. They can challenge us to look at our practice in a different way, or 'as' a different thing. They also give us a framework for simplifying, deconstructing, analysing and reconstructing our practice. Finally, they provide a framework for generating discussion and research.

As integrated practitioners we give ourselves a better chance of being more 'expert' if we can become familiar with, understand and experiment with the different models. The more we do that, and the more models we use, the broader and deeper our understanding of the reality is likely to become. Even if we disagree with particular models, simply by entering into a dialectical relationship with them we can challenge and either confirm or adapt our existing beliefs, ideas and practices.

THE MODEL AS A TYRANT

However, a model is always that: a model. It is a simplified version of reality, drawn down and abstracted, usually for the point of practising, teaching or analysing.

Nobody would want to take to the air in a model plane, nor to the sea on a model boat.

But because 'models' usually come down from on high, through teachers or academic departments, and because they often form the basis of curricula, assessments and exams, they may assume an overvalued importance in our minds. Overvaluing is a form of imbalance, and imbalance is a form of tyranny.

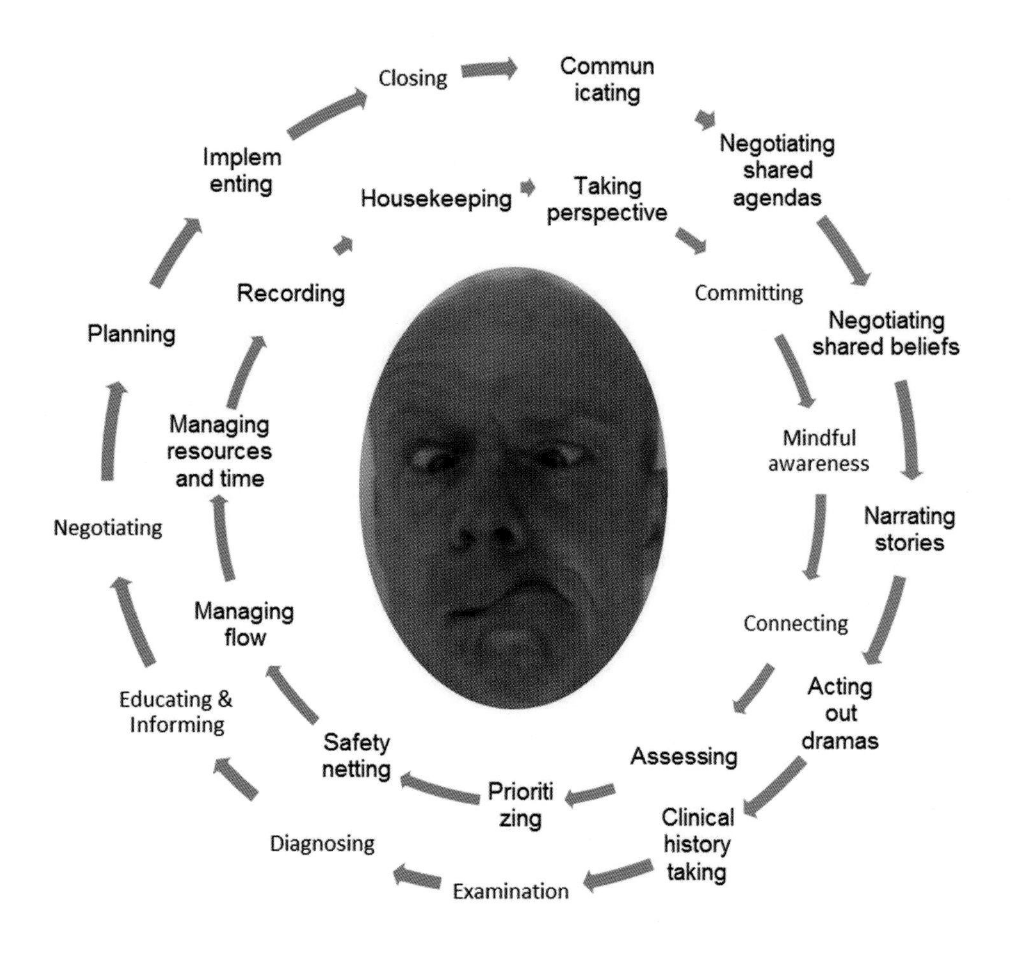

Here are some of the key ideas from the main medical consultation models dumped into one diagram. We can see that models can be seriously unhelpful, even tyrannical.

MODELS AS TOOLS OR TYRANTS?

There are many reasons why models may cease being tools and end up as tyrants.[53]

- They may be too numerous, so we get bewildered and confused.
- They may be too narrow, so we lose sight of the big picture.
- They may be too simplistic, so they fail to take account of the fundamentally chaotic and complex nature of our actual practice.
- They may be too time-structured, so we forget the crucial point that practice always happens in the moment, sometimes chaotically and randomly, and does not necessarily flow from a to b (or even from b to a).
- They may be too positivist, so that we fall into the trap of considering patient, practitioner and practice as independent, concrete objects for analysis; and so fail to take into account of the fundamentally relational and contextual nature of our existence.
- They may be too dualistic, so that we get the illusory sense that there is a fundamental difference and separation between patient and practitioner, or between assessment and management, or between causes and outcomes, and so fail to take into account that these things coexist and are co-created in a fundamentally interconnected, inter-relational and 'whole' experience of existence.

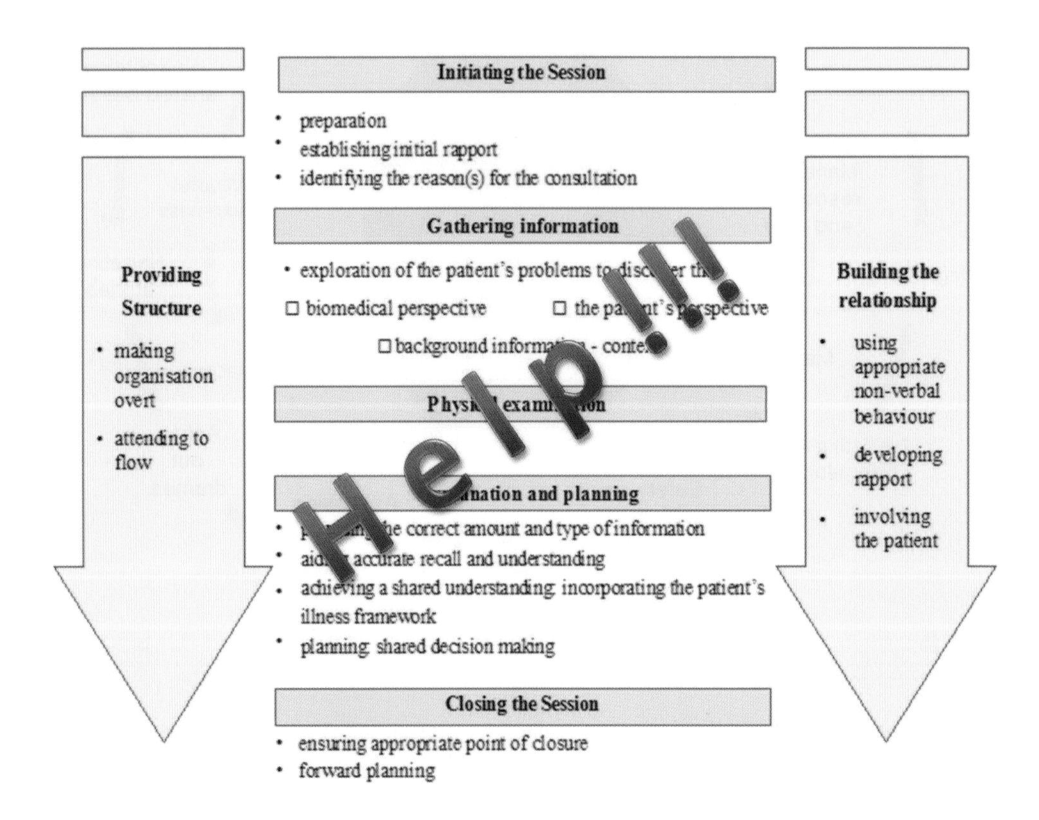

This ('Calgary Cambridge'[54]) consultation model is one of the most helpful of all, and the one I use most frequently in teaching and training. But I can quite safely say none of my consultations has ever looked anything like it. And those straight, bold, remorseless arrows really freak me out.

Models, maps, lines and circles give the impression of sequential, one-directional order and control, which can be helpful if we feel out of control in practice; but they can be too controlling when we want to imagine, create and express new realities within and through our practice.

INTEGRATING AND BALANCING MODELS AND PRACTICE

If our practice feels chaotic and out of control, abstracting reality into theoretical models is a very helpful thing to do. It gives us time and space to play around with the ideas and models in our heads, where we can make mistakes and learn without harm.

However, if we try to work the other way round, abstracting reality from theoretical models, we will come unstuck. If everything is related to everything else, attempts to divide off one or other part, perspective or relationship may be useful for learning and analysis, but will also ultimately be distorting. Reality is far more contextual, more messy, and more random (or at least chaotic) than we can ever capture in a model. Let's remember that models are our tools, not the other way round.

If you see the Buddha in the road, kill him.

– Zen Master Linji

In the same way, let us take care not to let our models become our gods. And if we do, and if we come across them in our actual practice, while we may feel that killing is a bit too violent, we may at least nod kindly in their direction, and wave them gently off.

So, for the rest of this book, we are going to say goodbye to models, in an attempt to see what we can discover by looking at our practice as it actually is in our experience: messy, relational and, at times, chaotic.

In this book so far, we have done exactly that by separating out the 'me', 'we' and 'other' relationships. In practice, of course, these three relationships happen at one and the same time, in an integrated way, within an integrated whole, which is the overall and integrated experience of existing. We have even, briefly, suggested a new model – the 'No-model Model'.

Like most things, it won't last.

I can't apply one rule to other models, and another to my own, so it's going to have to go too (painful though it is to release one's own creation).

With apologies to all of the wonderful authors, researchers and teachers who have helped me and countless others. Of course we still need you, only not actually 'in' the consultation. So, TTFN.

Bye-bye model. This one is SOOO much harder to let go. 'Vanity, vanity, all is vanity'[55]

THE REALITIES OF PRACTICE

In the last few chapters, we have talked about various things such as awakening, clearing, contacting, mapping, thinking and creating. These may sound suspiciously like 'tasks'. When we put them in order, it may look suspiciously like a 'process'. And when we present them all as a book section, it might look suspiciously like a 'model'.

So, for example, we could make them look like this (the Amery infinite-random-task-process-aka-bulldust model).

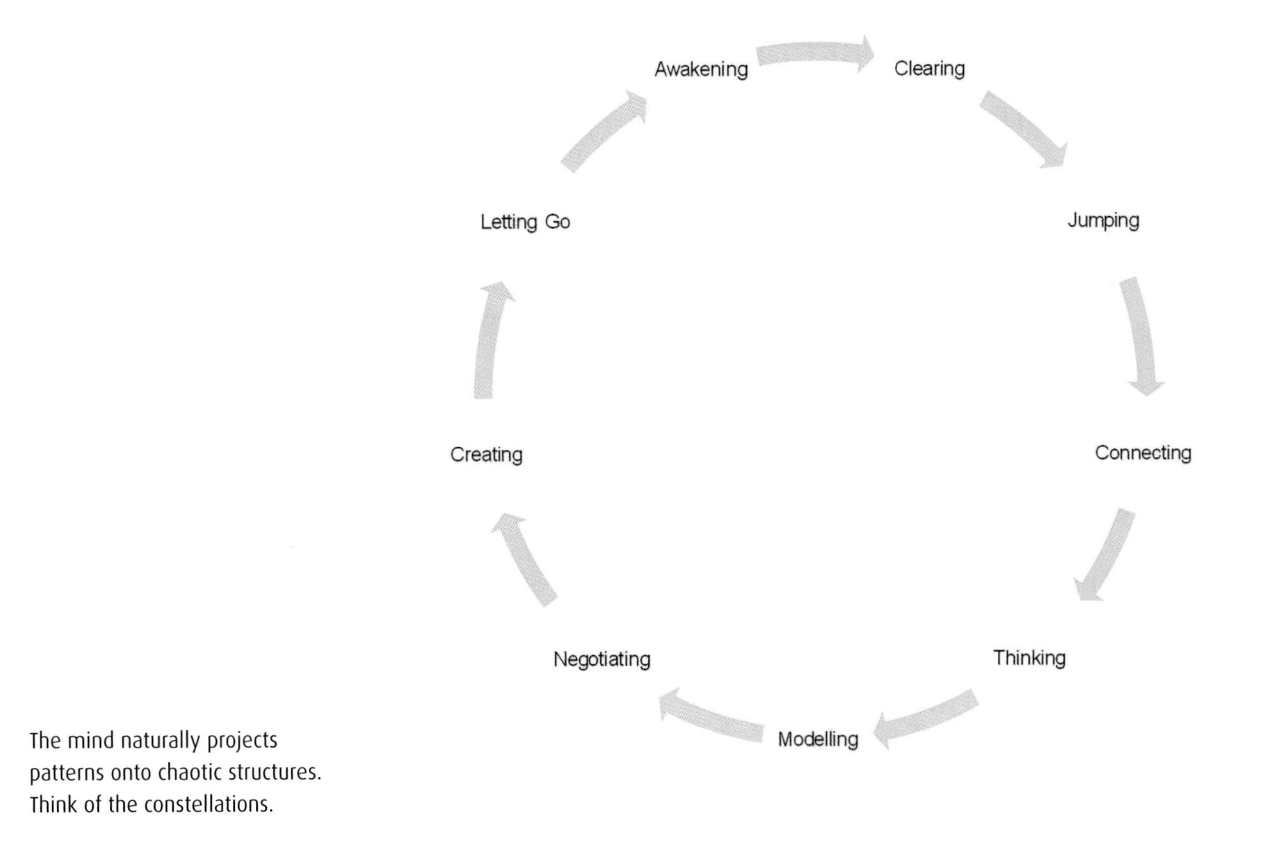

The mind naturally projects patterns onto chaotic structures. Think of the constellations.

And, in one way, they are. We are not suggesting that there aren't certain tasks that recur in health practice. We are not saying that these tasks don't often occur in a certain order, or process. And we are not saying that we can't model what happens in practice.

But we are saying that, in an infinite world, anything and everything can happen. But things don't always happen. Some things happen that we don't expect. Some things arise which are quite bizarre. When things do happen, they don't always happen in a particular process or order. And models are not the territory.

So for example, 'my' practice may also look like this . . .

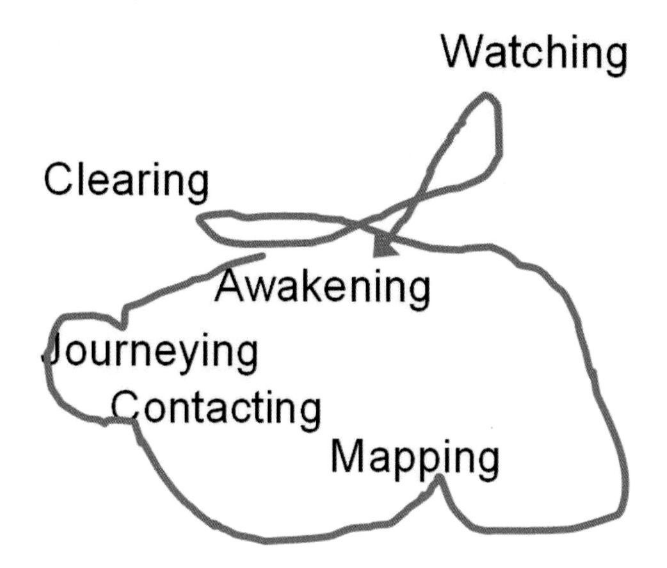

Or this . . .

Or even this . . .

AND NOW FOR SOMETHING COMPLETELY DIFFERENT

In these books we have divided things up into chapters and put these chapters in an order. But that is not to mean that, in practice, these are the 'right' chapters, or that there are no other 'better' chapters that have been written. It doesn't mean that, in practice, we have to follow them in this order or in any order at all.

Each chapter stands alone, and falls alone. Please use them to create your own practice as you wish. Or not. The universe is crazier and incorrigibly more plural than we might think.

Incorrigibly Plural

Incorrigibly plural
Louis gazes skyward
Peeling and portioning the world
As snow
Dancing drunkenly in the rose sky
Defying description
Touches him suddenly
Now

– JA[56]

Activity 13.2: Bye-Bye Buddhas (30 minutes)

Sit quietly and reflect on your Buddhas. Who are the people you most wish to emulate? What models do you particularly aspire to follow? What systems and practices do you most hope to use?

Try and visualise them clearly in your mind. As you visualise and re-experience them, take note of any feelings or emotions that arise. Do they comfort you? Do they give you a sense of belonging and purpose?

Now, in your mind's eye, let them go. You don't need to kill them. There's no call for violence here. But just let them drift away. Smile at them and wave. Perhaps even shed a mental tear. But let them disappear over the horizon.

Become aware of yourself, of yourself as a practitioner. See yourself with your patient. No models, no systems, no Buddhas. Just you, the patient, and the infinite 'other'.

Allow any feelings or thoughts to arise and watch them as they pass.

When your mind is clearer, become aware of a giant, blank canvas set up before you. Around you, as far as the eye can see, is an infinite palette of colour. In your hand is a paintbrush.

What can you create?

Chapter 14

Creating better health

Activity 14.1: Integrating everything and creating something (30 minutes)

Think of your most awesome or your most awful moment in practice, when you felt most alive, most in awe, most attuned to everything around you. It may have been a moment of joy: perhaps the time you delivered your first baby, or saw your first patient, or made your first incision, or when you received your first, heartfelt 'thank you' from a patient. It may have been a moment of pain: your first death, your first big mistake, your first angry patient.

What was it that made that moment of practice so awesome or awful? And what is it that makes other moments of practice so much less awesome, or so much less awful?

What, or rather who, is the common link between them all?

Reflect again on everything we have been through in these workbooks: on your relationship with yourself; on whether your tyrants are your tools (or vice versa); on how you communicate and co-create with your patients.

This is your practice.

Now.

What are you creating?

Can you improve that creation?

If so, how?

So here we are, at the point of creation.

We have our perspective. We are committed, aware and awake. We have cleared our minds and emotions and connected with our patients. We have mapped and planned our journeys. We have recognised emptiness and let go, so that we can fill the moment fully. We have engaged our minds, consciously and subconsciously, so that we can think fast and we can think slow. We have used our models wisely, but they are now waved off, because they can no longer help.

It is time to create what we create: better health.

CREATING IN YOUR OWN PRACTICE

I don't know what kind of health practitioner you are, what your training and background are, or which kind of practice you prefer. I don't know about your personality, beliefs, training, values, habits, neuroses, skills or competencies. I don't know what tools you have available to you, or how some of these tools might tyrannise you. I don't know anything about your patients, who they are, why they might choose to come to you, what they believe, what they hope for or what they expect.

In these books I have nothing more to give you except to encourage you to remember that, right now, you are creating. You and I are co-creating. Somehow the universe has configured itself so that you have become conscious of me and I have become conscious of you.

This self-consciousness and consciousness of others is a continuous and ongoing creative act. All experience is self-created and co-created.

So, as a health practitioner, you are fundamentally creative. You may also be scientific, scholastic, technical, analytical, empirical, logical or psychological too, but that doesn't stop you being creative. There may be times where you feel like just another cog in the wheel, that you have no control, perhaps even that there is no point. These perceptions are creations too.

Whatever you are, you cannot stop creating, because being conscious is being continuously creative. In every moment of our practice, we are creating and co-creating health. So the question is not: can I be a health practitioner and still be creative? The question is, how skilfully creative can I be in my practice of creating better health?

In this book we have also argued that better health comes from more harmonically balanced integration of all the various 'me', 'we' and 'other' factors in our patients' lives and in our lives. If that assertion is correct, it follows that the more harmonically balanced and integrated we can be through our practice, the better chance we have of creating better health.

How that creation comes out will depend on you, your patients and the tools you have available. All that remains is to decide whether you want to create more or less competently, more or less skilfully.

This is it.

Integrated practice in action

Activity 14.2: Integrating and creating (1 week)

Next time you are in practice, try pulling everything together, and integrating it into one whole creation.

At the beginning of the day, take perspective. Are you aware of your weakness or focusing on your strengths? Are you feeling tyrannised or can you see the many useful tools at your fingertips? Do you see your patients as needy and exhausting; or as fellow-travellers on a journey of co-creation? Does the space in front of you look tight, confined and unwelcoming; or does it look open, free and inviting? Is your practice a hassle and frustration, or is it what you love, what you are good at, perhaps even part of who you are?

Remind yourself of your values, rededicate yourself to what is important to you, and commit yourself to the moment.

Prepare as much as you can: get your environment, your tools, your equipment, your mind and your body as prepared as you can be. Allow enough time and space for this.

Before you start, stop: each day, each session, each consultation and each contact. Clear your mind, relax your body, wave off any tyrants, and allow your ego and its concerns to dissolve.

Let go. Trust yourself.

For a moment, an instant, touch the void, the infinite moment between jumping and not jumping. Experience Nothing.

Connect with your patient. Allow the flood of information to wash over you, but not to sweep you away. If emotions or thoughts emerge, don't engage, but watch them and take note of them. They will come in handy.

Listen deeply: not just with your ears, but with your full being. Don't interrupt the patient. Don't interrupt yourself. Give your subconscious time and space to work.

Watch as things start to arise: thoughts, feelings, emotions, sensations. Consider: whose are they? Feel what it is like to be the patient. Feel what it is like to be you.

When the flood has passed, think. Start to re-engage your conscious mind, but mindfully. Don't close out your subconscious, but allow your rationality and practicality to take over.

What are going to co-create? Visualise it. Experience it.

The blank canvas is before you. Feel the joy of allowing your expertise and experience to take over: checking, planning, modelling, negotiating, creating and co-creating. It doesn't matter if you are a surgeon or a physician, a doctor or a nurse, traditional or alternative, professional or amateur. Your skills are useful. Use them now. Make the idea into reality.

Create. Be created.

This is it.

Chapter 15
Integrating everything

Activity 15.1: Something and nothing (15 minutes)

In practice we are often called upon to reflect.

At the risk of inflaming you, please briefly reflect upon reflection.

Does it help you in practice? If so, how and when? If not, why not?

Please bear with me, but reflect even more. Who is doing the reflecting, and who is being reflected upon?

Where are they?

What is the space between them?

Before we draw everything to a conclusion, and come up with one or two suggestions for the future, let's just take a breather. Why don't we take a few minutes to reflect?

In fact, let's reflect on reflection.

Such is the nature of consciousness that, whenever we do something, we are also watching ourselves doing it (and also watching ourselves watching ourselves doing it . . .).

Consciousness is a continuous reflective process of self-perception and self-creation

CYCLING ON THE ROAD TO NOWHERE

In 1983, Donald Schön wrote the *Reflective Practitioner*. One of its central themes was the challenge to practitioners of all types to see their work not just in terms of technical competence but also in terms of artistic expression, requiring both to achieve excellence.

His theories centred on the concept of learning through reflection. This of course is a fundamental process of nature, even at cellular level: sense – react – sense. It is therefore a fundamental process of us.

Schön highlighted that, as self-aware beings, we don't just learn by reflecting 'in' our actions (i.e. reacting skilfully to events that we experience during the event itself); we also learn by reflecting 'on' our actions. This is an abstract, 'after the event' reflection that we can use to think forward to similar possible scenarios and adjust how we will behave as and when they arise in the future.

His work was challenging to previous models of how we develop 'excellence', which suggested that excellence comes from the steady and linear acquisition and application of a pool of expert knowledge.

He was not the first[57] to discuss this reflective and cyclical nature of learning, nor did he suggest that knowledge acquisition was unimportant. But he emphasised that being a practitioner is about being an artist as well as being a technician. He suggested that, as well as studying and accumulating knowledge, we also need to reflect and learn new and creative ways of application of that knowledge by reflecting on our experiences and practice.

This work led to a whole succession of cycles, for example Kolb's learning cycle,[58] Gibbs' supervision cycle,[59] and Rolfe's reflective model.[60] These have led, in turn, to many new models and tools for professional practice which have been tremendously effective, for example supervision, audit, significant event analysis, colleague feedback, patient feedback, reflective diaries, and 'PUNS and DENS'.[61]

Which is all very well. But we have gone back to geometric shapes again. Cycles are certainly more reflective than lines, but they are still lines, only joined up.

Cycling tightrope – cycles are arguably more interesting than straight lines and arrows, but do they actually get us anywhere?

THE MESSINESS OF LIFE

As we have tried to suggest in these books, the universe isn't really like that. It can't be reduced to neat, geometric shapes, although it can contain and express neat, geometric shapes. But it is also messy, infinite, interconnected, complex to the point of chaos, and absolutely relational.

Of course we can imagine and project onto our existence lines, circles, triangles, yin–yang and infinity symbols. But these are abstractions, and hence they are simplifications.

That's not to say that there is anything inherently wrong with simplifications or abstractions. Our consciousness is both. Sometimes it can be very helpful to abstract out, simplify and focus on one part of existence, separated from all the other parts. That's how we develop and perfect specific new skills and new ways of looking at existence.

However, like all tools, simplifications can become potentially tyrannical if they begin to dominate our learning, practice and teaching.

Very often in practice, things don't work out according to the models, or the patterns, or the guidelines. At such times, because the models and patterns and guidelines tend to have been handed down to us by 'the great and the good', it is easy to think there is something wrong with our practice. There may well be something wrong with our practice. And it may be that models and patterns can help.

But it may also be that the model, or the pattern, or the guideline, simply doesn't 'fit' the messy reality of this particular patient, sitting in front of this particular practitioner, at this particular point on the space–time continuum.

If, in our practice, we note that things seem to happen in fairly ordered and predictable patterns or sequences, that's great. Pattern recognition is a very effective and efficient tool in health practice, and one which develops continuously with experience.

But being a practitioner means keeping a broader perspective and being more mindfully aware, because sometimes things happen apparently randomly, chaotically, or in reverse.

If we are to be integrated practitioners, as well as reflective practitioners, we need to able to manage, and integrate it all: order and chaos, fullness and emptiness, reflection and action.

BEGINNER'S MIND: META-COMPETENCE

Being complex and chaotic, the universe will occasionally and randomly throw up situations that throw us. They may be situations that seem to lie outside of the guidelines, have no targets, or contradict the models. They may be situations that we have no experience of, in which case no amount of reflective practice will help us, as we have nothing relevant to reflect upon. Or they may be situations that we are simply unqualified to deal with, because we don't have the knowledge or skills.

'See through the eyes of a child' may seem like a rather trite saying. But just hold on a minute. Compared to older brains like ours (well, mine anyway) a baby's brain is like a super-computer logged on to super-fast broadband. They perceive and process everything, have no preconceptions, and judge as they find, uncoloured by years of sedimentary prejudice and ignorance. What's not to like?

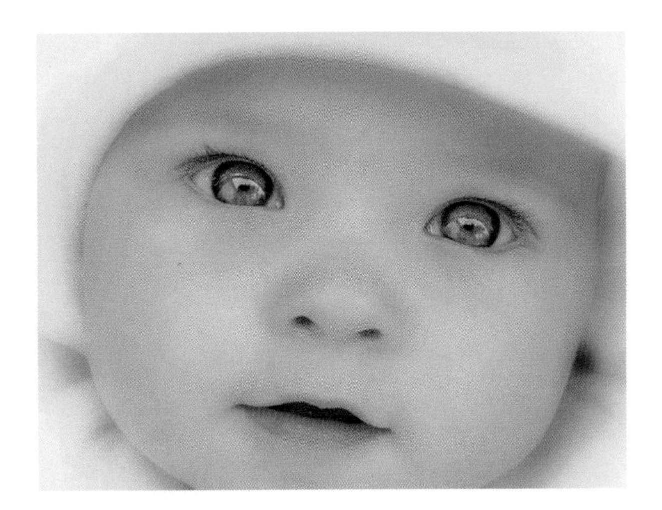

In the face of such complexity and randomness, we are like children. We are meeting things for the first time. We are empty.

The good news is children can teach us something here, because they are comfortable with emptiness and happy to keep themselves open, rather than looking at the world through any particular shaped spectacles. No preconceptions.

That's not to say we should be like children. Children are what is known as 'pre-competent'. They don't have knowledge and expertise, and they have not yet been taught how to reflect systematically on or use that expertise.

Having expertise, and the expertise to use that expertise, is called being 'competent'. We have got to the stage of expertise where we can call ourselves 'health practitioners'. That's no easy task, and so we should remember to give ourselves a quiet and unobtrusive but nevertheless heartfelt pat on the back. We can reflect on what we do, on our expertise, and through our reflections we can become even more expert. That's reflective practice.

But there is another step, called being 'meta-competent'. That means recognising we have expertise, and that we have expertise in using that expertise, but then deliberately and mindfully zooming further out, emptying ourselves, so that we stay aware of our expertise but also so we can see what is happening around us with un-blinkered eyes.

Look at this picture. It's me (again . . .)

Here I am, Super Mario ('Mario-complex'?). Hovering just above my head are my four 'super-powers'. These are four types of knowledge: factual knowledge, skills, experience and reflective ability.

Each of these types of knowledge is a meta-knowledge of the former (knowledge about how to use knowledge).[62] I can learn facts, I can put that factual knowledge into practice using my skills, I can develop and use my skills more naturally and effectively with experience, and I can improve all of these by reflecting on successes and failures.

At any one moment I can be explicitly aware of any or all of these types of knowledge; or they can be metaphorically hovering in the air, ready for me to grab when I need them. I can draw on factual knowledge. I can recall and use learnt skills. I can draw on experience when things take a turn for the worse. And I can also, at any time, go 'meta', go up a level, and watch myself using my knowledge, skills and experience; and then watch myself watching myself, and then reflect on myself watching myself watching myself, and then . . . OK, you get the picture.

But, just at this moment, I am not using any of them. I am not even looking at them. I know they are there if I need them. Right now I am just quietly aware of what is going on inside and out: clear, poised, and ready for action.[63]

Empty.

INCLUDING BUT TRANSCENDING OUR EXPERTISE

The theory behind 'meta-competence' is that we become more expert in our expertise the more we use it; so we become ever more effective and efficient in using it.

When we first start to practice, we are aware of our expertise but we use it very deliberatively and clumsily, like a learner driver clunking through the gears and jerking the brakes.

With practice we start to use our expertise intuitively, becoming aware of answers which are arrived at almost subconsciously, like when we change gears and brake consciously but smoothly.

Eventually, we become so effective that we react to situations almost automatically. But it is not that we are automated. It is that we have *included* but *transcended* the expertise, to become as aware as a child, all over again. Like when we drive along, enjoying the view, planning lunch, and singing 'La Traviata' all at the same time.

In the same way, as we become more meta-expert in the application of our expertise, we become more 'meta-aware' of our awareness. Initially we are very conscious of what we are doing, and that absorbs all our attention. With practice, we begin to be able to 'zoom out', taking in a bigger and bigger picture. Initially, we need to actively think about or discuss what we are doing to make sense of it. Later, we find that we

When Johann Sebastian Bach was asked how he found his melodies, he answered: 'The problem is not about finding them, it's – when getting up in the morning and getting out of bed – not stepping on them.'

are becoming 'instantly reflective' – reflecting on what we do at the same moment we do it. Eventually, we can almost dissociate, emptying ourselves and watching ourselves practice as we are practicing; being aware of our awareness as we are aware.

In simple terms, if we try and keep all our knowledge and expertise in the forefront of our 'minds', we will be distracted by them, and less able to be aware of, and focus on, what is happening right here, right now. If we can keep our various superpowers within reach but out-of-sight, we can really watch the universe with open eyes, confident we have tools to use as and when we need them.

INTEGRATED PRACTICE AS A 'META-LEVEL' OF REFLECTIVE PRACTICE

We are not suggesting that reflective practice is somehow 'wrong', just as reflective practitioners didn't suggest technical theory-based practice was 'wrong'. The former is a meta-level of the latter. Reflective practice is a meta-level of technical practice.

In the same way we are considering that integrated practice is a meta-level of reflective practice. It is a state of continuous total presence in every moment; and *at the same time*, a state of being continuously totally aware that we are continuously totally present.

Integrated practice is not about consciously 'finding' the 'problems' and 'solutions' (reflectively or otherwise). It's about becoming aware that the problems and solutions

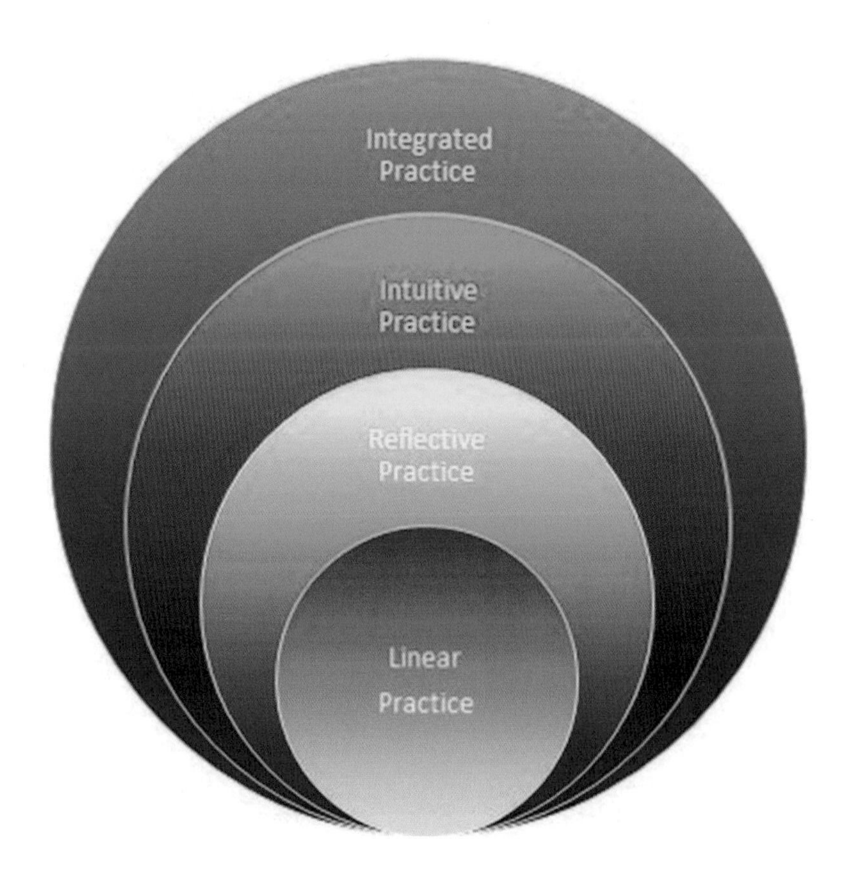

Each level includes and transcends previous levels.

will find us, and that we will find the tools we need to meet them, as long as we concentrate on not stepping on the problems and solutions because we are too busy 'remembering', 'reflecting' or 'acting'.

If we think about it, this is not at all odd. When we drive, we don't think: 'I am driving. I must think about how I am driving, and then make sure I drive according to Blogg's model or Smith's cycle.' We just drive. We might even enjoy the view of the universe passing by. We are calm and poised. Then, when something suddenly runs out into the road, we are alert enough to stop.

ZEN AND THE ART OF HEALTH PRACTICE

In practice, we are in the hugely privileged and exciting position of being able to watch and partake in the full range of human existence and experience. We can be artists, scientists, technicians, politicians, teachers, learners, even poets. We can support the beginning of life, the end of it, and every stage and manifestation of it in between. We get every opportunity to be alive, and every opportunity to throw ourselves into life, whatever it has to offer and in whichever creation it comes.

But we also get the opportunity to watch. And, in watching, we allow ourselves to become aware of the universe, watch as the universe expresses itself through us, and be transformed as the universe shapes itself through us. The universe creates as it expresses, and expresses as it creates. As an integrated part of the universe, the same applies: we create as we express, and as we express ourselves we create ourselves.

As we watch the universe express and create us, we become more aware of the absolute relationality and interconnectedness of all things. We become aware that, somehow, every now and again, the universe is able to balance things so perfectly, so harmonically, that 'I' appear, and with the appearance of the 'I', comes 'you', and 'them', and everything else.

As integrated practitioners we are aware that it is that balance, that harmony, which we aim to maintain, restore and make whole. That is healing. Healing is not a special skill. It is a skill that is open to all humanity. But it is when we combine it with the skills, knowledge and experience of health practice that we become able to be health practitioners who are 'good enough', and perhaps even a little bit better than that.

'The Abstract Kiss' – by Veronica Jackson:[64] a jumble of random shapes or a couple kissing. What are you creating?

The Kiss

Naked, open and empty.
Deeply, wholly free.
What is left, what is left,
What is left of me?

I was, I thought, I am.
I am, I am, am I?
Am . . . perhaps. But I?
Who on earth am I?

As space expands, and time rolls out,
Perhaps they sometimes kiss.
The faintest touch, the slightest breath,
A moment of infinite bliss.

And in that touch my soul appears.
A tiny spark set free.
A sign, a beacon, pointing to
The deep eternity.

Eternal peace, eternal joy,
Profoundest clarity.
To be, to see, then to return.
That, I think, is me.

– JA

Activity 15.2: Including and transcending in practice (30 minutes)

Reflect through these workbooks. Try to think of everything you have read. The 'me' stuff, the 'other' stuff, the 'we' stuff, the 'integrating everything' stuff. Every last thing.

We can't. It's not possible.

Does that mean it has gone? That our time and effort has been wasted? Reflect on that.

Reflect further back. Through your whole training and career. Try to think of everything. The knowledge, the information, the skills, the applications, the networks, the connections. Every last thing.

Chances are, you've forgotten more than a hundred times what you have remembered.

Does that mean it has gone? That our time and effort has been wasted? Reflect on that too.

Stop reflecting.

Seriously, stop it.

Breathe.

Let go.

Trust yourself.

It has not gone. If you keep your mind clear it will be there when you need it. You have transcended it.

You are now free, to be present, to express yourself, to create.

Welcome to integrated practice.

Conclusion: power, beauty and love

In these workbooks we have tried to dance and play around one central question: how can we be integrated practitioners in the 21st century?

We started from the premise that we practise with all of ourselves: our hands, ears, eyes, words, thoughts, values, beliefs, hopes and even dreams. We have discussed how our practice can be many things: science, technology, art, magic and even a thing of beauty. As practitioners, we hope to be able to integrate and balance all of these strands.

We have recognised that there are many entities in ourselves and in our practice that can tyrannise us, but we have also seen that we can regain mastery of these tyrants and start to use them as tools by subtly altering our perspective.

We have suggested that we are practising in an infinite world, with no clear start-points or end-points, and no map. This is unsettling even if things go well. And things don't always go well. Often we have to deal with stressful, painful or unpleasant things.

But the infinite, relational and messy nature of the universe turns out to be a blessing in disguise, for it gives us the power to express and create ourselves, and our patients, in ways that are better integrated, more harmonically balanced and more whole. In other words: more healthy.

We have discussed how integrated practice involves the balanced and harmonic expression of ourselves into and through the universe. In return it also involves us allowing the balanced and harmonic expression of the universe into and through us.

Through these expressions we are enabled to create and to make whole our own existences, and from there we have the opportunity to co-create and heal our patients, in a way which resonates with the whole of creation, and in a way that allows creation to re-create us.

As practitioners, we have more power than we might think. From the universal and basic building blocks of the universe we are almost miraculously able to create the full richness, complexity, texture, depth and colour of our existence. It's not that we are not gods, or martyrs. But perhaps we have indeed 'eaten from the tree of knowledge' and become aware of ourselves, as selves. Perhaps we have indeed eaten from the 'tree of life' and, as a consequence, been given a consciousness that is able to create itself, its life.

It seems we are, at our core, absolutely creative. And so in these books we have tried to place this creativity in its rightful place, back at the heart of health practice. It is not that we wish to turn the clock back. On the contrary, we value tremendously the advances that have come with the incorporation of science, technology and logistics into health practice. After all, these are also creations, and so they are also tools that we can and do use (with ever increasing skill).

But we have also chosen to remember that our practice is an art, because it is the creation of the self, and it is the creation of everything and everyone around us. If we can create in a way that is balanced, harmonic and integrated, we can create something that is 'whole' and 'healthy', which is something of beauty.

As practitioners we have to 'do'. It is not a question of simply sitting back and hoping things will sort themselves out. We have to think and choose and act. But as integrated and balanced practitioners we also have to 'be'. Perfection is unattainable. It is not a question of simply doing and expecting we can make everything right.

If we can find a way of both watching and practising, of both being and doing, we give ourselves the best chance of taking this messy, relational universe, and of taking this ordered, codified world of practice, and integrating all the diverse strands and sounds into a harmony, maybe even a symphony, within which everyone (ourselves included) feels healthier, happier and more whole.

Crafting creations that are happy, healthy and whole is ultimately an act of love. So perhaps we may end our journey with the thought that, while our practice may be many things, ultimately it is the practice of love, because it is the practice of creative and compassionate connection.

So, while we can think of health practice as many things, lets us remember also to think of it as the harmonic, balanced creation of beauty, through love.

Poetry, in fact.

Notes

1 The clue is in the title. Practitioners tend to be practical. While we might like to know the theory behind what we do, what tends to be more important is that it works. The original 'Integrated Practitioner' is a whole work comprising both theory and practice. This series of workbooks is intended to be more practical, so in workbooks 1–4 the practice will predominate. For those that are interested, the fifth workbook, *Food for Thought*, will discuss more of the theory that lies behind this work, as of course does the original book.

However, for now, please bear with us, as there are 13 key theoretical points that underpin this work and without which it may not make complete sense. They are as follows.

1. The universe, and every-'thing' within it, came into existence from no-'thing', and may presumably go back into nothing, and we can say nothing about the nothing, as there is nothing to say.
2. The universe and everything within it (including ourselves) is entirely and intrinsically relational. Within this relational web, certain states of matter and energy 'exist' (stand out) with varying degrees of complexity (entropy) against that background of nothingness.
3. Complex entities in the universe are holarchical. This means each level of complexity creates a whole which is greater than the sum of the parts. So, for example, clusters of atoms create molecules, clusters of molecules create cells, clusters of cells create organs, clusters of organs create beings, and clusters of beings create cultures and societies and biospheres. Each one of these can be said to exist on its own, as the interplay of smaller parts, and as part of the greater whole.
4. Fascinatingly, and slightly disturbingly, we find that things that may appear to us to be fixed are also relational. These include knowledge, truth, beliefs, meanings and eventually health itself. Not only are they relational, they are also self-referential. For example, truth is a function of meaning, meaning is a function of language, and language is a function of truth. Self-referential systems always end up in paradox. It is therefore impossible to define with certainty what 'health' is.
5. The universe is made up of the interplay between three things: forces, energy and matter. However, our experience of the universe is far, far richer than that. We feel warmth, beauty, taste, colour and texture. We experience anger, hope, fear, courage, joy and love. The reason that the universe appears so much richer to us is because of our consciousness. Consciousness takes in cold sense data derived from the forces, energy and matter of the universe, and uses them to

create the full richness of our existence. In other words, and in a very real way, our consciousness creates itself, and creates our experience of existence, as we go along.

6. While we think ourselves as having independent, concrete identity, this is actually just a matter of perspective. From a more macroscopic perspective, we are one infinitesimally small part of much larger relational systems: for example, our societies, our cultures, the biosphere, the noosphere, and the cosmos. From a microscopic perspective each one of our molecules and atoms comes from somewhere (or someone) else and goes somewhere (or to someone) else. From a quantum perspective we exist at the level of probability. From a cultural perspective the words, ideas and beliefs we use are mostly given to us by others.
7. When two conscious persons come into relationship with each other, each person's consciousness creates both itself and the other person. In other words, in relating to each other, in a very real way, we co-create each other.
8. Time does not flow. It is simply part of the space–time continuum. Our sense of time flowing derives from two things. First, our memory links together different states of existence in the space–time continuum in a linear way, giving us the idea that past flows into present. Second, our consciousness imagines future states of existence, giving us the idea that present flows into future.
9. This ability of consciousness to create past, present and future; to create itself; and to co-create others clearly has profound implications for what we think of as health, ill-health and health practice.
10. Health does not exist outside consciousness. It is a relational truth created by individuals, cultures and societies that has different meanings when viewed from different perspectives (for example, biomedical, psychological, sociological, or spiritual perspectives).
11. A common theme emerging from these different perspectives appears to be that health is something to do with the attainment and maintenance of a harmonic balance between different relational entities (for example, between molecules, between cells, between organs, between mind and body, between people, or between groups and societies).
12. While we cannot say what health is, we can suggest that health practice can therefore be seen as an attempt to co-create and maintain a harmonic, relational balance, not just for our patients but also for ourselves and our societies.
13. Being an integrated practitioner involves integrating all of the relationships and perspectives of our shared existence, using all of the tools that we have created and evolved through the history of human existence, to co-create 'healthier' states of existence from 'less healthy' states of existence. Health practice is therefore a science and a technology, but it is also fundamentally creative and therefore artistic.

That is enough of the theory. Let's get practical. After all, we are practitioners not theorists.

2 Edward Henry Potthast (1857–1927): 'Along the Mystic River'. Public domain art.

3 'Ars Poetica' by Archibald MacLeish, from *Collected Poems, 1917–1982*, Boston:

Houghton Mifflin; 1985. ISBN: 0395394171. Reprinted with kind permission of the Houghton Mifflin Company.

4 Open source, with many thanks to the poet and her family. The authors of the response are not known.

5 This image of an immune cell killing a cancer cell is from the Science Museum website www.sciencemuseum.org.uk as part of their wonderful 'Who am I?' online exhibition, with thanks.

6 Rollo May, *The Courage to Create* (May 1976).

7 This image is of a bowel anastomosis following traumatic injury. Image reproduced with thanks from www.trauma.org

8 Milton Glaser, in Lehrer 2012. Lehrer's book is well worth reading, and it makes some very interesting points for health practitioners, in particular that we are all creative. Creativity, and in particular brilliance in creativity, comes when a variety of threads can be woven together. First, there has to be the effort, the perspiration. In order to improvise, to head off the page, you need to know what's on the page first. Second, there is the need to create the right environment for creativity to work. Maximum creativity seems to happen when we ramp up the alpha waves: being relaxed, but not day dreaming. What we are aiming for is a state of mindful, relaxed awareness, peaceful but present. It's also not all right brain. We need the non-dominant hemisphere to generate openness and creativity, but we also need the left brain to shift those ideas into practical, doable works. We need divergent thinking to come up with ideas, and convergent thinking to make the creation happen. We need to be able to see the big picture, but also to focus on the intricacies of the small one. Perhaps the key thing is that we need to turn off the internal censor, forgetting about years of correction and humiliation, and take a different perspective. We are artists too.

9 Image reproduced under licence from www.photos.com

10 From a conscious perspective, it may feel as if intuitive creations and solutions appear out of the blue. However, as we will explore more in the final workbook, creativity can also be seen as the result of the interplay of various neurological processes: our subconscious working its way through our memorised store of previously recognised patterns and associations, looking for a pattern or association that matches the current problem, and suggesting new solutions based on old experiences.

11 Kahneman (2012) suggests that our intuitive thinking can be biased by a number of factors. If we can be aware of these factors, we can be more expert in using our intuition safely and effectively.
   - Always looking for a cause: our intuition wants to make connections, so it will look for causes even where the event was a chance event.
   - Mistaking plausibility for probability: something may happen. That makes it plausible. But it may be unlikely to happen. That makes it improbable.
   - Availability bias: the more emotion generated (e.g. anxiety) when we think about something, the more likely we think it is to occur, probably because our subconscious is making the possibility very 'available' to our intuition.
   - Risk aversion: people prefer to avoid loss than to acquire gains.

- Incorrect stereotype: for example, the notion that certain groups are more likely to be more or less prone to certain problems, when in fact there is no difference.
- Framing: the way a question is framed. For example, people are more likely to go for a treatment that has a 90% chance of succeeding than they will for one that has a 10% chance of failing.

12 Lehrer (2012) cites a number of experiments showing how our creativity and our ability to do hard mental reasoning can be affected by the environment. For example, our ability to look for, and find, creative new ideas and connections is assisted by helping our brains find cognitive ease: such as when taking a hot shower, going for a walk, having a drink or two, being around the colour blue. On the other hand, our ability to test the correctness of our ideas, and make them work in practice, is assisted by putting our brains under slight cognitive strain so that they can focus and stick at the task. For example, students asked to perform reasoning and short-term memory tests in red environments do better than in blue environments, whereas they do worse on imaginative or creative tasks. Lehrer also cites interesting examples of how different thinkers and creators use relaxants (such as alcohol or cannabis) to assist the generation of creative ideas, but use stimulants (such as caffeine or amphetamines) to give them the cognitive strength and resilience to make their creative ideas into reality. It is a delicate balancing act, though: too much relaxation simply leads to unproductive daydreaming, while too much stimulation leads to exhaustion and loss of perspective.

13 I have been unable to find the original source of this photo, which has been reproduced thousands of times on the internet.

14 The word to 'exist' comes from the Latin *ex* (out) and *istere* (to stand). So to exist means to stand out, which begs the question: from what? If, by definition again, everything that exists does exist, what does everything exist against? Nothing. So, without nothing could we not have something?

15 If so, great. Paradox is hard-wired into us, and indeed may be the force that drives consciousness (*see* workbook 5 if you'd like to explore this idea more).

16 NASA satellite photo of Hawaii.

17 The coastline is fractal. It is made up of wiggly bits, like peninsulas and coves. These peninsulas and coves are also made up of wiggly bits, which are themselves made up of wiggly bits. Each wiggly bit can be measured with the correct size ruler, although to measure the smaller and smaller wiggly bits, a smaller and smaller ruler is needed. The smaller the ruler, the more wiggles are measured, so the longer the distance. There is no theoretical limit to the shortness of a theoretical ruler, or to the number of wiggles within wiggles, so there is no limit to the length of the coast.

18 Our state of mind seems to be very important to creativity. If our thinking is too divergent, we may drift off into daydreaming, not arriving at any practical solutions. But if our thinking is too convergent, we may block out or miss possibly useful new ideas. Therefore, to be as effective as possible in practice, we need to allow time and space for both: divergence and convergence.

We can learn to use our brains in ways that make creativity more likely to happen. The main factor is the degree of 'cognitive ease' we feel. In general, factors that increase our sense of 'cognitive ease' will make us more likely to think divergently, whereas factors that increase our sense of 'cognitive threat' make us more likely to think convergently.

However, there are other factors too. Convergent thinking requires attention, information, stimulation, concentration and significant effort. Divergent thinking is facilitated by relaxation, quiet, a sense of well-being, a positive mood, and a positive, supportive, encouraging environment.

Knowing this, we can create conditions to make our creativity more likely, fostering divergence or convergence, as the moment and issue dictates.

19 Creativity, and in particular brilliance in creativity, comes when a variety of threads can be woven together. (*See* note 8 above.)

20 'Ananda' by Ko Un (1997) Reprinted from *Beyond Self: 108 Korean Zen Poems* (1997) by Ko Un with permission of Parallax Press, Berkeley, California, www.parallax.org.

21 'From 'The Collected Poems of Robert Creeley, 1945–1975', Robert Creeley (Author). University of California Press, 2006. ISBN: 9780520241589. Reprinted with kind permission of UC Press.

22 By Melvin Oliver and James 'Trummy' Young, and later covered by Fun Boy Three and The Bangles.

23 Information on many of the relaxation techniques can be found at mindtools.com. There are many forms of contemplative prayer, but a popular one at the moment is 'centring prayer' developed and described by Thomas Keating (www.centeringprayer.com). The 'whoosh' technique, a neurolinguistic programming technique, is a way of visualising a rapid, even instantaneous, clearing of the mind. An example can be found at www.planetnlp.com

24 'Constantly risking absurdity', by Lawrence Ferlinghetti (from *Coney Island of the Mind*) New Directions Publishing (1 February 1968), reprinted with permission of New Directions Publishing.

25 Lehrer (2012) quotes a number of studies showing how we are pretty sure we are about to solve a problem even before we have actually solved it. This is a very strange phenomenon. How do we know we are going to solve something unless we have actually solved it? The answer seems to be that the brain is constantly working ahead and calculating the probability of success of certain trains of thought, and it seems to be quite effective at it. It even boosts us with warm, feel-good sensations when we are on the right track, perhaps to encourage us to keep going.

26 For memory to be successful and adaptive it needs to include the ability to acquire, store, and retrieve relevant memories. Memories that are acquired in traumatic or frightening situations tend to be very easily accessible, for obvious survival reasons. In post-traumatic stress disorder, memories of the trauma remain so 'accessible' that they can recur without warning, in vivid detail, as 'flashbacks'. This also gives a clue to another important function of memory, and that is 'memory extinction'. If we have too many easily accessible memories, our minds will become cluttered as we react to one flashback after another. Somehow, the brain manages to segregate memories into those that are most contextually useful, which are kept easily accessible and experienced vividly, and those which are less accessible and only remembered cognitively (i.e. we can remember that the event happened, but we

don't get all the sounds, sights, smells and sensations that originally went with the event when it comes to our mind). We can learn to use our memory as a tool, by allowing ourselves to experience moments as they are, and segregating out those for which we only want cognitive memories. This is why, as health practitioners, we are able to experience extremely traumatic situations without becoming overwhelmed by the memories of them.

27 Shunryu Suzuki-roshi was a Buddhist master who started to bring Zen teachings to the West in the last century. He wrote: 'The practice of Zen mind is beginner's mind. The innocence of the first enquiry – what am I? – is needed throughout Zen practice. The mind of the beginner is empty, free of the habits of the expert, ready to accept, to doubt, and open to all the possibilities. It is the kind of mind that can see things as they are, which step by step and in a flash can realise the original nature of everything.' (Suzuki and Dixon 1970)

28 Displayed currently in the Tate Modern in London and reproduced with the kind permission of the Estate of Roy Lichtenstein: © Estate of Roy Lichtenstein/DACS 2013.

29 Doctors tend to leave an average of only 18 seconds for the patients to talk at the beginning of the consultation before they interrupt; and we interrupt around 70% of our patients before they have finished explaining why they came to see us. When we do interrupt, only a tiny fraction of patients manage to get back to complete the information they had planned to give us. This is a big problem, because patients do not always (or even often) tell us their biggest concern first. They will often start with an 'opening gambit', and keep their major concern back until they feel more calm and trusting. The longer we wait before interruption, the more complaints are elicited and 'late arising' (just one more thing doctor . . .') is significantly reduced. We also tend not to ask our patients what their ideas, concerns and understandings are (only 6% in one study) yet failure to elicit patients' ideas, concerns and expectations leads to poor understanding, poor adherence, poor satisfaction and poor outcomes. (Schegloff, Jefferson & Sacks 1977; West 1979; Barsky 1981; Frankel 1984; Beckman & Frankel 1984; Tuckett *et al.* 1985)

30 In practical use, we use the word 'think' to apply to such a broad range of different activities that we probably have to accept that the idea of 'thought' is rather a meaningless but convenient label for the almost infinite array of our mental processes; for example, reasoning, imagining, dreaming, creating, destroying, wondering, memorising, inducing, judging, deducing, believing, predicting, translating or even simply being aware.

31 The fastest form of thinking is divergent thinking, sometimes called 'intuition', 'right brain thinking' or 'system one thinking' (Kahneman 2012). It is very quick, creative, subconscious, sensitive to patterns and changes in patterns, and almost effortless. It seems to 'work' by spotting and creating associations and patterns between existing bits of knowledge. It can be misleading, but in the right circumstances can be very effective and efficient. We tend to switch to divergent thinking when we are relaxed, at ease, and have time and space for our minds to wander freely. The safest form of thinking is convergent thinking, sometimes called 'reasoning', 'system one thinking' or 'left-brain thinking'. It is much slower and harder work to carry out, and

requires complete focus to be effective, so we get tired quickly when we do it. It is very sensitive to specific states and builds on known premises to develop logical connections and deductions, infer, hypothesise, model and create new knowledge. We tend to switch to this when we have a task to complete, when we are under pressure or if we feel under threat.

As an example, look at the Müller-Lyer illusion below. Which line do you think is longer?

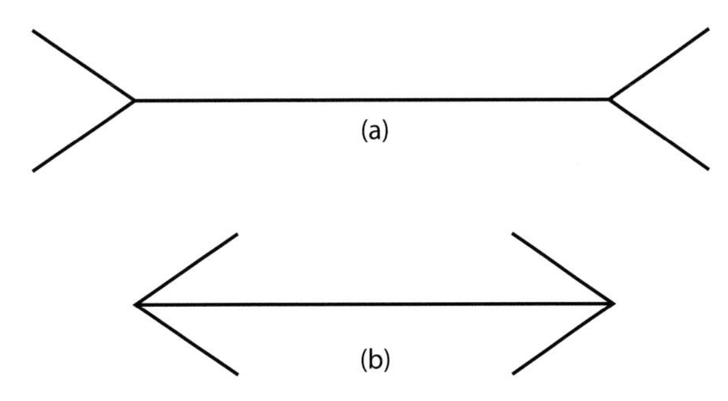

If you have not seen this illusion before, the chances are your intuition will say it is 'a'. You can correct your assumption logically, by taking out a ruler and measuring it, but that's much harder work.

32 It was the second one. Unless you are a horse racing expert and recognise the colours, the chances are your intuition would have been no more accurate than guessing.

33 Kahneman (2012) suggests we will intuit answers to problems even when we are completely new to the problem. As novices, our intuition is much more likely to be wrong (even though we intuitively feel it is right). This is maybe why we are nervous of using intuition in practice. However, he also demonstrates that intuition becomes safer and more effective as we become more experienced and familiar with the situation. This is much more like the situation that prevails in our own health practice – in which by definition we are more expert and familiar than others. It seems we may be avoiding using a tool which is fast, effective and not tiring to use.

34 The middle-aged chap (no. 1) is having an MI. No. 2 has anxiety-related chest pain and no. 3 has indigestion. If you have been a health practitioner for any length of time, you probably found this one a lot easier than test one (and you probably got there without even 'thinking' – although of course you were thinking, just subconsciously and intuitively). Photos under license from www.photos.com

35 Kahneman (2012) suggests that our intuitive thinking can be biased by a number of factors. If we can be aware of these factors, we can be more expert in using our intuition safely and effectively.

- Always looking for a cause: our intuition wants to make connections, so will look for causes even where the event was a chance event.
- Mistaking plausibility for probability: something may happen. That makes it plausible. But it may be unlikely to happen. That makes it improbable.
- Availability bias: the more emotion generated (e.g. anxiety) when we think about something, the more likely we think it is to occur, probably because our subconscious is making the possibility very 'available' to our intuition.
- Risk aversion: people prefer to avoid loss than to acquire gains.

- Incorrect stereotype: for example, that certain groups are more likely to be more or less prone to certain problems, when in fact there is no difference.
- Framing: the way a question is framed. For example, people are more likely to go for a treatment that has a 90% chance of succeeding than they will for one that has a 10% chance of failing.

36 Lehrer (2012) cites a number of experiments showing how our creativity and our ability to do hard mental reasoning can be affected by the environment. (*See* note 12 above.)

37 Traditionally, we tend to split the process that practitioners and patients undertake as linear, often dualistic, processes. For example, 'assessment/treatment'. However, 'assessment', treatment', 'patient' and 'practitioner' are relational concepts. In practice, things are much more messy, chaotic and contextual than that. Our minds are limited in how much information and how many linkages they can hold at any one moment, so we sometimes need to extract information and model it in a simplified way. That is fine, as long as we don't fall into the trap of thinking the model is the reality.

38 This is a tricky point to grasp. Thinking of our existence as in some way 'not real' is an inversion of the modern way of thinking 'I think therefore I am'. The obscure philosophy does not matter too much. however. We will go into it more in workbook 5.

39 Copyright of the Pollock-Krasner Foundation ARS, NY and DACS 2013.

40 Different practitioners will also have their own maps, and the breadth and depth of these maps will vary from practitioner to practitioner. For example, a surgeon is primarily concerned with the anatomy of their patient, a geneticist with the genetics. These are hugely important roles. Without specialisation we would never have achieved so much in healthcare. Therefore we are not equating breadth with 'good' practice and depth with 'bad' practice. Some more maps might be as follows. These are suggestions only, because of course only the individual drawing the map knows what is in his or her map.

General physician

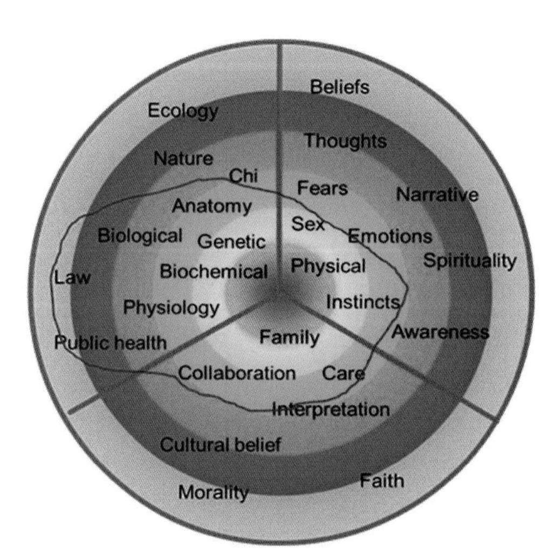

Psychotherapist

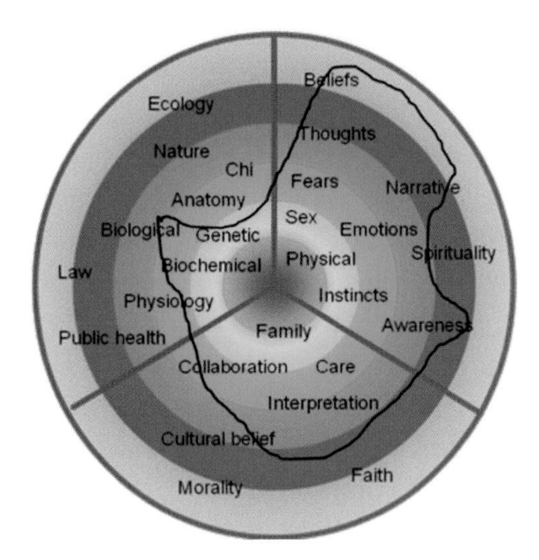

Traditional healer

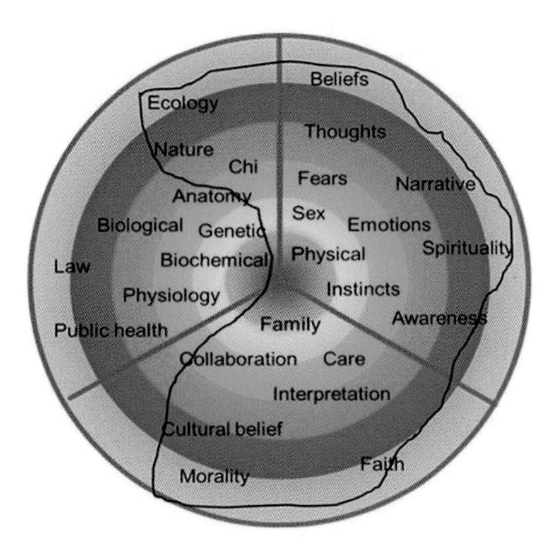

41 We looked at why this might be in workbook 3. We all give off many, many signals to each other all the time: the way we dress, the words and phrases we use, the metaphors and imagery. It may be a slight feeling of disconnect, visual or verbal clues and cues that suggest we are not seeing eye to eye. It may be transference: we may feel ourselves getting irritated or frustrated, or we may feel them getting irritable or frustrated with us.

42 The wished for outcomes and the wished for processes in any health journey are known as the 'agendas' of patient and practitioner. These wishes are based upon the perspective we take on our own health, and on health practices that we believe may be effective. In turn these perspectives are based upon personality, social class, ethnic group, how vulnerable or threatened we feel, an estimation of the beliefs of treatment weighed against cost, risks and inconvenience; and trigger factors, such as alarming symptoms, advice from family or friends, messages from the mass media, disruption of work or play.

We have already discussed in workbook 3 the importance of understanding the patient's concerns and hopes. As patients may hide their 'real agenda' at the start of the consultation, it is really important to look for it. This is not just a question of 'practising nicely', important though that is. It is also a question or practising quickly and effectively. Failure to elicit the patient's full agenda early leads to longer consultations, wrong paths, blind alleys, inappropriate, expensive and time-wasting examinations, investigations and treatment. On the other hand, if we can discover early the patient's whole agenda, we give ourselves a better chance of planning and pacing our consultation (and ourselves) and allocating sufficient time and effort to things like discussing, educating, planning and carrying through. The most simple way to establish the agenda is to try to create a trusting and open rapport as soon as we meet, not interrupting (so the patient completely finishes his or her opening remarks), verbally or non-verbally valuing each problem as it comes out, but not engaging with any problem until the whole list is out on the table. We might use variations of a theme along the lines of: 'OK, that's interesting, and we will certainly cover that if you'd like. But before we do is there anything else that you'd like to discuss/that is on your mind/that you were concerned about/that's at the back of your mind?' Once we are sure the whole agenda is out, we can agree a priority list with the patient. Quite often, during this prioritisation, we find that the patient decides he or she does not really want to address the lesser concerns and so they may disappear off the agenda altogether, thus saving even more time. The gentle but dogged persistence in getting the agenda out early seems, in my experience, to be particularly hard for trainees to pick up, usually as the trainee is anxious that, by asking for more, they will be overwhelmed by an enormous problem list and no time left to deal with it. In fact, paradoxically, the opposite is most often the case.

43 'Boaters Rowing on the Yerres', Gustave Caillebotte. 1877–79. Public domain art.

44 Helman's folk model of illness (Helman 1981) suggested some questions that patients tend to ask themselves before they seek the advice of a health practitioner.
- What has happened?
- Why has it happened?
- Why to me?
- Why now?
- What would happen if nothing were done about it?
- What should I do about it?

45 For example, doctors may want to explore symptoms and signs, run tests, attempting to 'diagnose' an underlying 'pathology', and select the right 'treatment'. Nurses may hope to help the patient achieve a fuller expression of their needs, or to empower the patient to care for herself, or to act as a guide to better health, or to model more healthy behaviour. Hypnotherapists may wish to achieve a deeper understanding of the patient's fears, thought processes and emotions, look for subconscious blocks and drivers, and influence more healthy decision making and behaviour.

46 With thanks to www.gptraining.net

47 Summarising and signposting help by clarifying for both patient and practitioner where we have got to and what our choices of route still are. They also provide an opportunity for both patient and practitioner to 'come out of role' briefly and check in with each other that things are still going OK. In that sense they are collaborative, and lead to a more powerful partnership. They also can prevent either patient

or practitioner feeling a sense of loss of control. By summarising, our patients learn that we are interested in them and taking notice of them, as well as confirming to them that we have understood. For the practitioner, it keeps us from going off down blind alleys (and so wasting time), helps us get our thoughts together and raise flags if we have got to a problematic issue.

Some useful signposting and summarising phrases might include the following.

Starting off:
- So?
- Where should we start?
- Why don't we begin by . . .?

Switching
- Have we finished talking about that?
- Do you think we should move onto . . .?
- We've considered X and Y and Z; which of these do you think we should move on to?
- Would it be helpful if we looked at this a different way?
- Now, turning to . . .

Generating more discussion
- Well, I wonder where all that leads us?
- Might it be helpful if we consider this in more detail?
- I wonder what all this means?

Adding texture and context
- This reminds me of the time . . .
- A good example of this might be . . .
- To illustrate this point . . .

To suggest alternatives
- I can see that, but another perspective might be . . .
- Let's consider a hypothetical situation for a moment. If you were to . . .
- Would you mind if I played the devil's advocate for a moment? What if . . .

Ending
- So, to conclude . . .
- Shall we sum up?
- Perhaps you could recap?
- Shall we summarise what we've covered?

48 Poem in the public domain.

49 This painting is entitled '*Loss: Mary and Dead Christ*' by Jennifer Lamontagne, www.jenniferlamontagne.com/contact.html and reproduced with kind permission of the artist.

50 There is a number of different consultation models, which can be summarised below (in chronological order).
- 1957: *The Doctor, His Patient, and the Illness*. The Balints had psychoanalytical training and applied this understanding to the consultation. This entirely novel approach demonstrated that the consultation can be therapeutic in itself, despite

what the practitioner does, and demonstrated that the practitioner was not a neutral observer but an active participant in a relationship with the patient.

- 1964: *Games People Play* (doctors talking to patients). Eric Berne pointed out that we are fundamentally 'transactional' entities, continuously in transactions with ourselves through our internal discourses and relationships. (*See* Chapter 9 'Acting' in workbook 2.)
- 1975: 'A six-category intervention analysis'. John Heron suggested in more detail the kinds of ways practitioners can use consultations to the benefit of the patient. These include being prescriptive (advising or instructing), informing, confronting (challenging maladaptive behaviour or attitudes), catharsis (the therapeutic release of emotion), catalysing (encouraging and supporting the exploration of subconscious ideas and concerns) and supporting (comfort, approval and affirmation).
- 1976: *Doctors Talking to Patients* by Byrne and Long. They were the first to analyse the structure of the consultation, and using tape recordings they suggested that consultations may occupy a spectrum from 'doctor-centred' to 'patient-centred' and that consultations tend to follow a similar pattern and structure (establishing relationships, discovering reasons for attendance; a history and possibly examination; consideration of the condition; treatment and/or further investigations; closure).
- 1979: 'The exceptional potential in each primary care consultation' by Stott and Davis was the first to look at the kinds of themes that may be covered in consultations. For example: managing the presenting problem, modifying behaviours, managing continuing problems and health promotion.
- 1981: 'Disease versus illness in general practice'. Helman's anthropological background led him to suggest some explanatory models that need addressing in the consultation. These include attempting to discover the answers to six questions.
  - What has happened?
  - Why has it happened?
  - Why to me?
  - Why now?
  - What would happen if nothing were done about it?
  - What should I do about it?
- 1984: *The Consultation*. Pendleton, Schofield, Tate and Havelock suggested that there are seven 'tasks' that need to be completed in a consultation, which include the following:
  - defining the reason for the patient's attendance (and in particular they highlighted the importance of discovering not just the overt history but also the more hidden 'ideas, concerns and expectations' of the patient)
  - considering other problems
  - agreeing appropriate actions for each problem
  - achieving a shared understanding of the problems with the patient
  - involving the patient in their own management and responsibility
  - using time and resources appropriately
  - establishing and maintaining a relationship with the patient.

- 1987: *The Inner Consultation* by Roger Neighbour explored the combination of practitioner–patient relationship within the context of a consultation and suggested that there are five parts to the relationship:
  - connecting (developing rapport and empathy)
  - summarising (reasons they came and also ideas, concerns and expectations)
  - 'handing over' (suggesting, explaining and agreeing a management plan which is then taken on by the patient)
  - 'safety-netting' (scenario planning to cover worst-case scenarios)
  - 'housekeeping' (where the practitioner becomes aware of and reacts effectively to his or her own needs and emotions deriving from practice in order to stay healthy).
- 1989: Three Function Approach to the Medical Interview (which is the model adopted by The American Academy on Physician and Patient). It suggests that consultations have three functions requiring various different tasks to be completed:
  - gathering data: using open-ended questions, facilitation, checking back, negotiating, clarifying, directing, summarising, eliciting ideas and expectations, assessing impact
  - developing rapport and responding to patient's emotion: using reflection, legitimation, support, partnership and respect
  - education and motivation: educating about illness, negotiating and maintaining a plan, and motivating patients to adhere to the plan.
- 1997: *Communicating with Medical Patients*. Stewart and Roter suggested that consultations follow two parallel pathways which need to be recognised and integrated to result in shared understanding, planning, decision making and action:
  - the 'illness framework' (usually the patient's agenda)
  - the 'disease framework' (the doctor's agenda).
- 2000: The Calgary Cambridge method uses comprehensive analysis of evidence to suggest an integrated model which incorporates both 'tasks' of the consultation (similar to Pendleton *et al.*) but also the skills required to run the consultation effectively (in particular 'providing structure' by commentating and signposting what is happening for the patient; and 'maintaining relationship' by using effective communication and empathic skills).
- 2002: *Narrative-based Primary Care* by John Launer, applying the principles of narrative analysis and therapy to the consultation – *see* Chapter 5 'Storytelling' in workbook 2 for more detail.
- 2002: *Consulting with NLP* by Lewis Walker is an attempt to bring neurolinguistic programming techniques into the consultation.

51 Just like doctors, nursing theorists have suggested various models for nursing consultations and practice. These include the following.
- Grand theories: which focus on and emphasise certain desirable behaviours for nursing practitioners.
- Comfort theories: which suggests the role of nursing is to provide and extend comfort to the patient and family, so encouraging healthy behaviours and attitudes.

- Adaptation theory: which suggests that the nurse acts as a guide, helping and supporting patients as they move through the ever-changing nature of existence and illness and thereby enabling the patient to feel more comfortable and confident with their situation.
- Developmental theory: which suggests that the nurse assists the patient as he or she develops along life from conception through to death.
- General systems theory: which suggests that the nurse acts as someone who acts upon the whole system of the patient's existence, and intervening to discover and manage any issues that might compromise the effectiveness and health of the whole system.
- Humanistic theory: which looks at nurses and patients as human beings in relationship with each other and the world; and in which the nurse is able to act as an observer and commentator, questioning, challenging and supporting the patient. To be effective, the nurse has to have self-awareness and awareness of the greater context of both the nurse's and the patient's existence in the broadest sense.
- Modelling and role-modelling theory: a social learning theory within which the nurse aims to understand the patient's explanatory and beliefs models and behaviours; and then acts to support effective beliefs and attitudes and model better beliefs and behaviours where necessary, with the aim of helping patients become stronger, more positive and empowered.

52 Some of these have been covered in previous chapters. However, we have not yet mentioned gestalt therapy, which is surprising as there are strong parallels between the approach of gestalt therapy and the approach of this book, in particular the relational nature of the self, the experience of self phenomenologically and ontologically, and the co-creation of the moment between relating people.

53 Models can be very useful tools, particularly for teaching and learning. However, they can become tyrants in certain circumstances, as in the following examples.
- Too many: there are now so many models, generating even more tasks, viewpoints, attitudes and processes, it is practically impossible for anyone to put them all into practice. Models that are unrealistic are unbalanced and therefore inherently tyrannical.
- Too narrowly focused: in the real world of health practice in busy health systems, there are many patient relationships that practitioners need to take into account during a patient consultation. These would include the patient in front of us, the patients waiting, the patients who need visiting, the patients who are unaware that we need to do something for/to them, the patient that just left, the patient who is not yet here but who may suddenly need our attention, and the patient who wrote a complaint about us last week. These patients do not usually appear in the consultation models, and yet they form part of the existence that practitioners create.
- Too geometric: people who write models seem to love geometric shapes. They may be linear 'task-based' models, circular 'process-based' models, triangular 'relationship' models. But geometric shapes do not exist. They are cognitive creations projected onto the universe. The universe is infinite four-dimensional 'foam'. As such, any shape can be superimposed upon it, and it will 'fit'. But the fit is one

way. My shape may fit into an infinite foam, but an infinite foam will not fit into my shape. Life is much messier than can be captured in shapes.

- Too chronological: events, like time, do not flow from past to present to future. They just are. Each state and each event is when and where it is. We can connect them in an order that is useful to us, but we must not forget we are doing the connecting. There is no actual connection (other than indirectly through the 4-D foam). It may be that the more states we experience, the more we notice patterns in those experiences. Things may happen in a certain order some of the time, or maybe even most of the time. However, in a consultation, states and events may not happen at all, or happen in a different order, or appear apparently quite randomly. Lines and circles give the impression of sequential, one-directional order and control, which can be helpful if we feel out of control; but can be too controlling when we want to create and express new realities.
- Too positivist: models may (with the exception of gestalt and transactional analysis models) separate the practitioner and patient out, dealing with them independently of each other and of their contexts, as if each is an autonomous, independent entity capable of separation and objectivity. In practice, practitioner and patient cannot be separated from the context of their coexistence, with each other or with the wider world.
- Too dualistic: models may appear dualistic; for example: patient-centred and practitioner-centred, patient's agenda and practitioner's agenda, patient's concerns and practitioner's concerns. Patients and practitioners are separate to an extent, but the key subject of health practice is their co-creation in relationship with each other. This co-creation is a single integrated whole, and it is this single integrated whole with which we have to work, and is the single integrated present upon which more or less healthy futures are built.

54 Kurtz, Silverman & Draper 2005.

55 Ecclesiastes Chapter 1, verse 2. The whole chapter goes like this:

> 1 The words of the Preacher, the son of David, king in Jerusalem. 2 Vanity of vanities, saith the Preacher, vanity of vanities; all is vanity. 3 What profit hath a man of all his labour which he taketh under the sun? 4 One generation passeth away, and another generation cometh: but the earth abideth for ever. 5 The sun also ariseth, and the sun goeth down, and hasteth to his place where he arose. 6 The wind goeth toward the south, and turneth about unto the north; it whirleth about continually, and the wind returneth again according to his circuits. 7 All the rivers run into the sea; yet the sea is not full: unto the place from whence the rivers come, thither they return again. 8 All things are full of labor; man cannot utter it: the eye is not satisfied with seeing, nor the ear filled with hearing. 9 The thing that hath been, it is that which shall be; and that which is done is that which shall be done: and there is no new thing under the sun. 10 Is there any thing whereof it may be said, See, this is new? it hath been already of old time, which was before us. 11 There is no remembrance of former things; neither shall there be any remembrance of things that are to come with those that shall come after.

56 With thanks (and apologies) to 'Snow' by Louis MacNeice, whose wonderful poem inspired this chapter.

57 Dewey suggested that we use reflective thought, which is an 'active, persistent, and careful consideration of any belief or supposed form of knowledge in the light of the grounds that support it and the further conclusions to which it tends' (Dewey 1910).

58 Kolb (1984) suggested we have different learning styles based around a learning cycle. The cycle of learning is:

- concrete experience (CE)
- reflective observation (RO)
- abstract conceptualisation (AC)
- active experimentation (AE)
- back to CE again.

These can be grouped into perceiving activities (experience and conceptualisation) and processing activities (observation and experimentation). Each of us tends to prefer a particular stage on the cycle, and therefore these govern our learning styles.

- Diverging 'Reflectors' (CE/RO) – prefer feeling and watching
- Assimilating 'Theorists' (AC/RO) – prefer watching and thinking
- Converging 'Pragmatists' (AC/AE) – prefer thinking and doing
- Accommodating 'Activists' (CE/AE) – prefer doing and feeling.

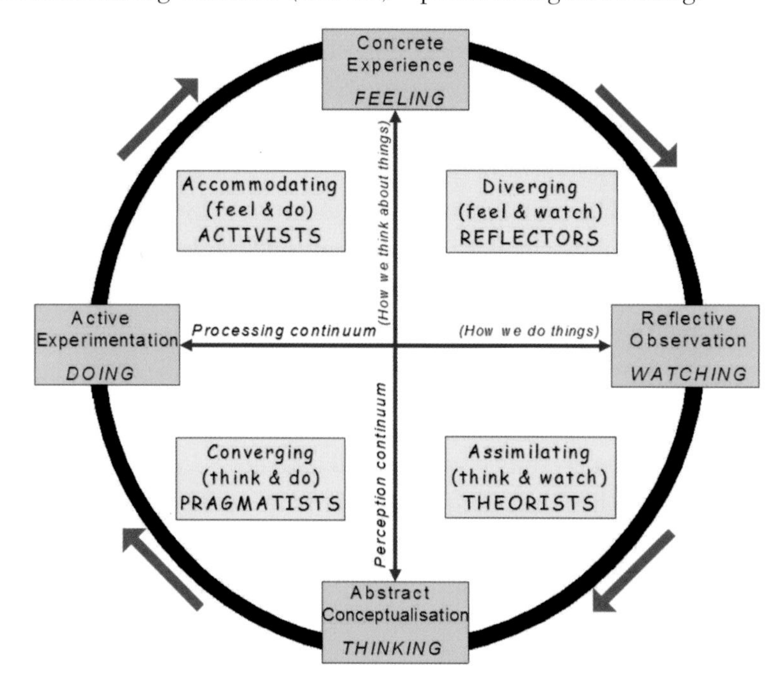

Diagram credit: www.brainboxx.co.uk

59 Gibbs' reflective cycle (Gibbs 1988), commonly used as the basis for clinical supervision, is helpful in that it provides a framework for breaking down different elements of practice and reflecting on them independently, before reintegrating them.

- Stage 1: Description of the event (facts only, no emotions or judgments).
- Stage 2: Feelings (here we can bring out the feelings and emotions that may have arisen).
- Stage 3: Evaluation (a discussion of what went well and less well).

- Stage 4: Analysis (a more detailed discussion of why things went well or badly, and what the various factors at play were).
- Stage 5: Conclusion (a decision about what went well and what would be done differently if it happened again).
- Stage 6: Action (exactly who will do, what, where, when and how if the same situation arises in the future).

60 Rolfe's (Rolfe, Freshwater & Jasper 2001) is similar but more simplified. It just has three stages.
- What: what happened, the effects, consequences, actions, responses, feelings, etc.
- So what: what it means, what has been learnt (individually and as a group/organisation), implications of relationships, attitudes, cultures and so on.
- Now what: specific actions that need to be done to improve future actions and outcomes.

61 There are numerous very helpful tools and templates for learning, training and appraisal which are applicable for practitioners everywhere. The Bradford Training Scheme website has a 'Trainers Toolkit' which is a veritable mine of useful stuff. Please look at www.bradfordvts.co.uk

62 According to Eraut (2000) practitioners tend to use a number of different processes in our work:
- assessing patients and situations
- deciding what action to take
- taking a course of action (and modifying this course as needed)
- meta-cognitive watching and monitoring of ourselves, others and the situation developing around us as we work.

As practitioners we also use three different modes of cognition, which we pick and choose depending on the urgency, speed and conditions required.
- Instant/Reflex: this is almost subconscious and automated.
- Rapid/Intuitive: which is conscious but only just, as we leap straight from observation to understanding and action without knowing how exactly we got there.
- Deliberative/Analytic: which is when we take in the whole picture and put the pieces together before arriving at a decision and action plan.

These processes and cognitive modes interact as in the table below.

Type of Process	Mode of Cognition		
	Instant/Reflex	Rapid/Intuitive	Deliberative/Analytic
Assessment of the situation	Pattern recognition	Rapid interpretation	Prolonged diagnosis Review with discussion and/or analysis
Decision making	Instant response	Intuitive	Deliberative with some analysis or discussion
Overt actions or action scripts	Routinised action	Routines punctuated by rapid decisions	Planned actions with periodic progress reviews
Meta-cognition	Situational awareness	Implicit monitoring Short, reactive reflections	Conscious monitoring of thought and activity Reflection for learning

As we acquire more experience and practice, tasks that required deliberative approaches gradually become more 'routinised' and so we become less aware of them. Thus this knowledge becomes 'tacit' rather than 'explicit'. Tacit knowledge is highly efficient as it gets us where we need to go quickly and effectively.

Tacit knowledge is not without problems. Because we are unaware of it operating, we may not recognise when situations and issues move on, so our short cuts become less and less relevant and secure. Therefore, in professional practice, tacit knowledge tends to be safety-netted by generating and testing hypotheses (e.g. diagnoses) or plans (e.g. management plans) against evidence (e.g. from clinical tests or published evidence) or the views of other people (e.g. colleagues or specialists).

The thing which distinguishes the 'expert' is not just what they know, but also how they use and apply what they know (because their knowledge has been continuously moulded and melded as a result of considerable experience).

The acquisition of tacit knowledge develops best in social contexts, where professionals can discuss and test knowledge with each other. It also requires an open and permissive culture of learning, wherein people feel able to share mistakes as well as successes.

Of particular relevance to health practice, time and thinking interact either in a constructive or destructive way. A certain degree of time shortage forces people to focus more attentively, process more quickly and adopt more intuitive practices. Similarly, a more intuitive approach frees us up to do things faster. More significant time limitations put limitations on meta-cognition; but a slightly less time pressured environment enables us to meta-process, thus taking into account such factors as self-awareness, awareness of the context and connectedness of problems, awareness of the various ways we can think about the problems, and selection of the most appropriate tools.

Very severe time limitations prevent us even using intuitive actions, as we are forced into almost entirely automated actions, with no time for meta-processing at all. Without some meta-processing, we rapidly lose confidence in our abilities, because we have no feedback against which we can check our actions and decisions. In this situation, the effectiveness and efficiency of our work drops off rapidly.

63 The importance of 'meta-cognition' is emphasised by Epstein (1999) in a fantastic paper published called 'Mindful practice'. He builds on Eraut, reinforcing the importance of meta-knowledge (also known as self-awareness or mindfulness) for the effective, efficient and expert practice. He suggests that mindfulness is the most important form of professional knowledge, as it 'informs all types of professionally relevant knowledge, including propositional facts, personal experiences, processes, and know-how, each of which may be tacit or explicit'. To be experts, we need to be able to control and apply both explicit and tacit knowledge. Experts 'use a variety of means to enhance their ability to engage in moment-to-moment self-monitoring, bring to consciousness their tacit personal knowledge and deeply held values, use peripheral vision and subsidiary awareness to become aware of new information and perspectives, and adopt curiosity in both ordinary and novel situations'. He contrasts the alternative, more common in junior and training practitioners, who are often so absorbed in the task itself they lose perspective and the ability to cross

check and safety net, thereby practising more slowly, less effectively, making more errors in judgement and technique. He suggests that 'although mindfulness cannot be taught explicitly, it can be modelled by mentors and cultivated in learners. As a link between relationship-centred care and evidence-based medicine, mindfulness should be considered a characteristic of good clinical practice.'

64 By Veronica Jackson, http://fineartamerica.com/featured/the-abstract-kiss-veronica-jackson.html. Reproduced by kind permission of the artist.

Bibliography

Abbasi K. Doctors: automatons, technicians, or knowledge brokers? *JRSM*. 2007; **100**(1): 1. Print.

Aked J, Marks N, Cordon C, Thompson S. Five ways to well-being. *Foresight Project on Mental Capital and Wellbeing*. New Economics Foundation; 2008. Web. Available at: www.neweconomics.org/publications/five-ways-well-being-evidence

Alladin A, Alibhai A. Cognitive hypnotherapy for depression: an empirical investigation. *IJCEH*. 2007; **55**(2): 147–66. Print.

Allen RP. *Scripts and Strategies in Hypnotherapy: the complete works*. Carmarthen: Crown House Publishing; 2004. Print.

Ambady N. Surgeons' tone of voice: a clue to malpractice history. *Surgery*. 2002; **132**(1): 5–9. Print.

Amery J. *Children's Palliative Care in Africa*. Oxford: Oxford University Press; 2009. Print.

Anielski M. *The Economics of Happiness: building genuine wealth*. Gabriola, BC: New Society; 2007. Print.

Armstrong D. Space and time in British general practice. *Soc Sci Med*. 1985; **20**(7): 659–66. Print.

Arnetz BB, Horte LG. Suicide patterns among physicians related to other academics as well as to the general populations: results from a national long-term prospective study and a retrospective study. *Acta Psychiatr Scand*. 1987; **75**(2): 139–43. Print.

Balint M. *The Doctor, His Patient, and the Illness*. New York: International Universities; 1957. Print.

Bandura A. Self-efficacy: toward a unifying theory of behavioral change. *Psychol Rev*. 1977; **84**(2): 191–215. Print.

Barsky AJ. Hidden reasons some patients visit doctors. *Ann Intern Med*. 1981; **94**: 492–8. Print.

Beating the Blues®. Web. Available at: www.beatingtheblues.co.uk (accessed 28 October 2011).

Beck DE, Cowan CC. *Spiral Dynamics*. Oxford: Blackwell; 2006. Print.

Beckman HB, Frankel RM. The effect of physician behavior on the collection of data. *Ann Intern Med*. 1984; **101**: 692–6. Print.

Beevers CG, Miller IW. Perfectionism, cognitive bias, and hopelessness as prospective predictors of suicidal ideation. *Suicide and Life-Threatening Behavior*. 2004; **34**(2): 126–37. Print.

Bench M. Open Door Coaching. Web. Available at: www.opendoorcoaching.com. (accessed 17 October 2011). Copyright © 2003 Marcia Bench and Career Coach Institute; reprinted with permission.

Berne E. *Games People Play: the psychology of human relationships*. New York: Grove; 1964. Print.

Betancourt JR, Ananeh-Firempong O. Not me! Doctors, decisions, and disparities in health care: how do we really make decisions? *Cardiovasc Rev Rep*. 2004; **25**(3): n.p. Print.

Better Health. Web. Available at: http://getbetterhealth.com (accessed 17 October 2011).

Black Dog Institute. *Depression*. Black Dog Institute. Web. Available at: www.blackdoginstitute.org.au (accessed 23 November 2011).

Blanck PD, Buck R, Rosenthal R. *Nonverbal Communication in the Clinical Context*. University Park: Pennsylvania State University Press; 1986. Print.

Blenkiron P. *Stories and Analogies in Cognitive Behavioural Therapy*. Oxford: Wiley Blackwell; 2010. Print.

Block N. How many concepts of consciousness? *Behavioral and Brain Sciences*. 1995; **18**(2):272–8. Print.

BMJ. How much do we know? Clinical Evidence. BMJ. Web. Available at: http://clinicalevidence.bmj.com/ceweb/about/knowledge.jsp (accessed 17 October 2011)

Bohm D. *Wholeness and the Implicate Order*. London: Routledge & Kegan Paul; 1981. Print.

Bradford VTS. Trainers' Toolkit. Home. Web. Available at: www.bradfordvts.co.uk (accessed 12 November 2011).

Brantley J. *Calming Your Anxious Mind: how mindfulness and compassion can free you from anxiety, fear, and panic*. Oakland, CA: New Harbinger Publications; 2007. Print.

British Association for Behavioural & Cognitive Psychotherapies. Home Page. Web. Available at: www.babcp.com (accessed 28 October 2011).

British Medical Association. *Doctors' Health*. 8 May 2007. Web. Available at: www.bma.org.uk/doctors_health/doctorshealth.jsp?page=2 (accessed 28 October 2011).

British Medical Association. *Quality and Outcomes Framework, February 2010*. Web. Available at: www.bma.org.uk/employmentandcontracts/independent_contractors/quality_outcomes_framework/qualityframework10.jsp (accessed 28 October 2011).

Brown D. Evidence-based hypnotherapy for asthma: a critical review. *IJCEH*. 2007; **55**(2): 220–49. Print.

Bruton HJ. Book review: nations and households in economic growth: essays in honor of Moses Abramovitz (Paul A. David, Melvin W. Reder). *Economic Development and Cultural Change*. 1979; **27**(4): 801. Print.

Bstan-'dzin-rgya-mtsho, Hopkins J. *Becoming Enlightened*. New York: Atria; 2009. Print.

Buber M. *I and Thou*. New York: Continuum; 2004. Print.

Buchbinder SB, Wilson M, Melick CF. Estimates of costs of primary care physician turnover. *Am J Managed Care*. 1999; **5**(11): 1431. Print.

Businessballs. *Job Satisfaction Inventory*. Businessballs Free Online Learning for Careers, Work, Management, Business Training and Education. Web. Available at: http://businessballs.com (accessed 27 October 2011).

Businessballs. Web. Available at: http://businessballs.com (accessed 24 October 2011).

Byrne PS, Long BEL. *Doctors Talking to Patients*. London: HMSO; 1978. Print.

Campbell DT. Blind variation and selective retention in creative thought as in other knowledge processes. *Psychol Rev.* 1960; **67**: 380–400. Print.

Campling P, Haigh R. *Therapeutic Communities: past, present, and future*. London: Jessica Kingsley; 1999. Print.

Campo R. What the body told. *The World in Us: lesbian and gay poetry of the next wave*. New York: Griffin; 2001. N.p. Print.

Caplan F, Caplan T. *The Power of Play*. New York: Doubleday; 1973. Print.

Carroll L, Green RL. *Alice's Adventures in Wonderland; and, through the looking-glass and what Alice found there*. London: Oxford University Press; 1971. Print.

Casey PR, Tyrer P. Personality disorder and psychiatric illness in general practice. *Br J Psychiatry.* 1990; **156**(2): 261–5. Print.

Chomsky N. A minimalist program for linguistic theory. *The View from the Building: essays in honor of Sylvain Bromberger*. Cambridge: MIT; 1993. N.p. Print.

Cole SA, Bird J. *The Medical Interview: the three-function approach*. St. Louis: Mosby; 2000. Print.

Committee on the Use of Complementary and Alternative Medicine by the American Public. *Complementary and Alternative Medicine in the United States*. Washington, DC: National Academies; 2005. Print.

Covey, S. *The 7 Habits Of Highly Effective People*. Free Press; Revised edition 2004.

Cozens J. Doctors, their wellbeing and stress. *BMJ*. 2003; **326**: 670–1. Print.

Csikszentmihalyi M. *Finding Flow: the psychology of engagement with everyday life*. New York: Basic; 1997. Print.

Dalai Lama. *Becoming Enlightened*. London: Rider; 2010. Print.

Dalai Lama, Cutler HC. *The Art of Happiness: a handbook for living*. Audiobook CD. New York: Simon & Schuster Audio; 1998.

Dalai Lama, Hopkins J. *Becoming Enlightened*. New York: Atria; 2009. Print.

Damgaard-Mørch NL, Nielsen LJ, Uldwall SW. [Knowledge and perceptions of complementary and alternative medicine among medical students in Copenhagen]. [Article in Danish] Ugeskr Laeger. 2008; **170**(48): 3941–5. Available in translation at: www.vifab.dk/uk/statistics/medical+students+and+alternative+medicine?

Davison S. Principles of managing patients with personality disorder. *Adv Psychiatr Treat.* 2002; **8**: 1–9. Print.

Deber RB. What role do patients wish to play in treatment decision making? *Arch Intern Med.* 1996; **156**: 1414–20. Print.

de Girolamo G, Reich JH. *Epidemiology of Mental Disorders and Psychosocial Problems: personality disorders*. Geneva: World Health Organization; 1993. Print.

DeLongis A, Folkman S, Lazarus RS. The impact of daily stress on health and mood: psychological and social resources as mediators. *J Pers Soc Psychol.* 1988; **54**(3): 486–95. Print.

Dennett DC. *Consciousness Explained*. London: Penguin; 1993. Print.

Deveugele M, Derese A, van den Brink-Muinen A, *et al*. Consultation length in general practice: cross sectional study in six European countries. *BMJ*. 2002; **325**(7362): 472. Print.

Dewey J. *How We Think*. Boston: D.C. Heath & Co; 1910. Print.

Dickinson E, Franklin RW. *The Poems of Emily Dickinson*. Cambridge, MA: Belknap of Harvard University Press; 1998. Print.

Digman JM. Personality structure: emergence of the five-factor model. *Annu Rev Psychology.* 1990; **41**(1): 417–40. Print.

DiMatteo M, Robin CD, Sherbourne RD, *et al.* Physicians' characteristics influence patients' adherence to medical treatment: results from the Medical Outcomes Study. *Health Psychol*. 1993; **12**(2): 93–102. Print.

DOH. *Improving Access to Psychological Therapies (IAPT) Programme: computerised Cognitive Behavioural Therapy (cCBT) implementation guidance*. Department of Health, UK; March 2007. Web. Available at: www.dh.gov.uk/en/Publicationsandstatistics/Publications/PublicationsPolicyAndGuidance/DH_073470

DOH. *Delivering Care, Improving Outcomes for Patients.* Quality and Outcomes Framework; 8 February 2010.

DOH. *Mental Health and Ill Health in Doctors*. London: Crown Publishing; 2008. Department of Health. Web. Available at: www.dh.gov.uk/en/Publicationsandstatistics/Publications/PublicationsPolicyAndGuidance/DH_083066.

DOH. *Mental Health Policy Implementation Guide: adult acute inpatient care provision*. Department of Health (UK); 2002. Web. Available at: www.positive-options.com/news/downloads/DoH_-_Adult_Acute_In-patient_Care_Provision_-_2002.pdf.

DOH. *The GP Patient Survey: general information*. The GP Patient Survey. UK Department of Health; 2010. Web. Available at: www.gp-patient.co.uk/info

Doran T. Effect of financial incentives on incentivised and non-incentivised clinical activities: longitudinal analysis of data from the UK Quality and Outcomes Framework. *BMJ*. 2011; **342**: 590–8. Print.

Dowson JH, Grounds A. *Personality Disorders: recognition and clinical management*. Cambridge: Cambridge University Press; 1995. Print.

Dunnette MD, Hough LM, Triandis HC. *Handbook of Industrial and Organizational Psychology*. Palo Alto, CA: Consulting Psychologists; 1990. Print.

Durkheim É, Cladis CS. *The Elementary Forms of Religious Life*. Oxford: Oxford University Press; 2001. Print.

Durojave OC. Health screening: is it always worth doing? *The Internet Journal of Epidemiology*. 2009; **7**(1): n.p. Print.

Easterlin RA. Does economic growth improve the human lot? Some empirical evidence. In: David PA, Reder MW, editors. *Nations and Households in Economic Growth: essays in honor of Moses Abramovitz*. New York: Academic Press; 1974. Print.

Edelman GM, Mountcastle VB. *The Mindful Brain: cortical organization and the group-selective theory of higher brain function*. Cambridge: MIT; 1978. Print.

Edelman GM, Tononi G. *A Universe of Consciousness: how matter becomes imagination*. New York, NY: Basic; 2000. Print.

Ely JW, Osheroff JA, Ebell M. Analysis of questions asked by family doctors regarding patient care. *BMJ*. 1997; **319**: 358–61. Print.

Epstein RM. Mindful practice. *JAMA*. 1999; **292**(9): 833. Print.

Eraut M. Non-formal learning and tacit knowledge in professional work. *Br J Educ Psychol*. 2000; **70**(1): 113–36. Print.

Erickson HC, Tomlin EM, Price Swain MA. *Modeling and Role Modeling: a theory and paradigm for nursing*. Englewood Cliffs, NJ: Prentice-Hall; 1983. Print.

Ericsson KA. *The Cambridge Handbook of Expertise and Expert Performance*. Cambridge: Cambridge University Press; 2006. Print.

Ernst E. Obstacles to research in complementary and alternative medicine. *Med J Aust*. 2003; **179**(6): 279–80. Print.

Evans R. Releasing time to care: Productive Ward, survey results. *Nurs Times*. 2007; **10**(Suppl. 16): S6–9.

Eve R. *PUNs and DENs: discovering learning needs in general practice*. Oxford: Radcliffe Medical Press; 2003. Print.

Everett DL. *Don't Sleep, There Are Snakes: life and language in the Amazonian jungle*. New York: Pantheon; 2008. Print.

FearFighter. Panic & Phobia Treatment. CCBT Limited Healthcare online. Web. Available at: www.fearfighter.com

Festinger L. *A Theory of Cognitive Dissonance*. California: Stanford University Press; 1957. Print.

Figusch Z, editor. *From One-to-one Psychodrama to Large Group Socio-psychodrama: more writings from the arena of Brazilian psychodrama*. Figusch; 2009. Print.

Finke RA, Ward TB, Smith SM. *Creative Cognition: theory, research, and applications*. Cambridge, MA: MIT; 1996. Print.

Firth-Cozens J. Doctors, their wellbeing, and their stress. *BMJ*. 2003; **326**: 670–1. Print.

Flett G. York researcher finds that perfectionism can lead to imperfect health. *York's Daily Bulletin*. Toronto, Canada: York University; June 2004. Print.

Flood GD. *An Introduction to Hinduism*. New York, NY: Cambridge University Press; 1996. Print.

Flynn JR. *What Is Intelligence: beyond the Flynn Effect*. Expanded paperback ed. Cambridge: Cambridge University Press; 2009. Web. http://en.wikipedia.org/wiki/International_Standard_Book_Number

Foresight Project. *Mental Capital and Wellbeing: making the most of ourselves in the 21st century*. The Foresight Project. The Government Office for Science: London; 2008. Web.

Foucault M. *History of Madness*. London: Routledge; 2006. Print.

Fowler KA, Lilienfield SO, Patrick CJ. Detecting psychopathy from thin slices of behaviour. *Psychol Assess*. 2009; **21**: 68–78. Print.

Frackowiak RSJ, Ashburner JT, Penny WD *et al*. *Human Brain Function*. 2nd ed. San Diego, California: Academic Press; 2004. Print.

Frankel RM. From sentence to sequence: understanding the medical encounter through microinteractional analysis. *Discourse Processes*. 1984; **7**(2): 135–70. Print.

Fredrickson BL. The role of positive emotions in positive psychology: the broaden-and-build theory of positive emotions. *Am Psychol*. 2001; **56**(3): 218–26. Print.

Gabora L. The origin and evolution of culture and creativity. *Journal of Memetics*. 1997; **1**(1): n.p. Print.

Gardner, H. *Frames of Mind: The Theory of Multiple Intelligences*. 3rd ed. Basic Books, 2011. Print.

Gettier EL. Is justified true belief knowledge. *Analysis*. 1963. **23**: 121–3. Print.

Gibbs G. *Learning by Doing: a guide to teaching and learning methods*. [London]: FEU; 1988. Print.

Gilbert DT. *Stumbling on Happiness*. New York: Vintage; 2007. Print.

Gilbert E. *Eat, Pray, Love: one woman's search for everything*. New York: Penguin; 2006. Print.

Giles J. *No Self to Be Found: the search for personal identity*. Lanham: University of America; 1997. Print.

Gillon R. Medical ethics: 'four principles plus attention to scope'. *BMJ*. 1994; **309**: 184. Print.

Glaser BG, Strauss AS. *Awareness of Dying*. Chicago: Aldine Pub.; [1965]. Reprint 2005. Print.

GMC. *Disciplinary Decisions*. Rep. General Medical Council. Web. Available at: www.gmc-uk.org/concerns/hearings_and_decisions/fitness_to_practise_decisions.asp

GMC. *Good Medical Practice*. Rep. General Medical Council UK, 2006. Web. Available at: www.gmc-uk.org/guidance/good_medical_practice.asp

GMC. *Printable Documents*. Summer 2009. Web. Available at: www.gmc-uk.org/concerns/printable_documents.asp

Goldberg LR. The structure of phenotypic personality traits. *Am Psychol*. 1993; **48**: 26–34. Print.

GP Online. *A Registrar Survival Guide . . . setting up your consulting room*. GP Online. 2010. Web. Available at: www.gponline.com/Education/article/1037805/a-registrar-survival-guide-setting-consulting-room (accessed 4 November 2010).

GP Training Net. *Consultation Theory*. Web. Available at: http://gptraining.net (accessed 12 November 2011).

Grant J, Crawley J. *Transference and Projection: mirrors to the self*. Buckingham: Open University; 2002. Print.

Greene B. *The Elegant Universe: superstrings, hidden dimensions, and the quest for the ultimate theory*. London: Vintage; 2005. Print.

Greenhalgh T, Hurwitz B, editors. *Narrative Based Medicine: dialogue and discourse in clinical practice*. London: BMJ; 2002. Print.

Grimshaw GM, Stanton T. Tobacco cessation interventions for young people. *Cochrane Database Syst Rev.* 2006; **4**: CD003289. Print.

Haigh R. Modern milieux: therapeutic community solutions to acute ward problems. *The Psychiatrist*. 2002; **26**: 380–2. Print.

Haigh R. The quintessence of a therapeutic environment: five universal qualities. In: Campling P, Haigh R, editors. *Therapeutic Communities: past, present and future*. London: Jessica Kingsley; 1999. pp. 246–57. Print.

Hakeda YS. *Kukai: major works*. New York: Columbia University Press; 1972. Print.

Hall ET. *The Hidden Dimension*. Garden City, NY: Doubleday; 1966. Print.

Hammond DC. Review of the efficacy of clinical hypnosis with headaches and migraines. *IJCEH*. 2007; **55**(2): 207–19. Print.

Handy CB. *Gods of Management: the changing work of organizations*. New York: Oxford University Press; 1995. Print.

Handy CB. *Understanding Organisations*. Harmondsworth, Middlesex: Penguin; [1976] 1985. Print.

Hawking SW. *A Brief History of Time: from the big bang to black holes*. Toronto: Bantam; 1988. Print.

Health Foundation. *Evidence: helping people help themselves. A review of the evidence considering whether it is worthwhile to support self-management*. Health Foundation; May 2011. Web. Available at: www.health.org.uk/publications/evidence-helping-people-help-themselves

Health Talk Online. *Shared Decision Making*. Healthtalkonline. DOH. Web. Available at: www.healthtalkonline.org/Improving_health_care/shared_decision_making (accessed April 2011).

Hecht MA, LaFrance M. How (fast) can I help you? Tone of voice and telephone operator efficiency in interactions. *J Appl Soc Psychol*. 1995; **25**(23): 2086–98. Print.

Hélie S, Sun R. Incubation, insight, and creative problem solving: a unified theory and a connectionist model. *Psychol Rev*. 2010; **117**(3): 994–1024. Print.

Helman CG. Disease versus illness in general practice. *J R Coll Gen Pract*. 1981; **31**: 548–62. Print.

Hendrich A, Chow MP, Skierczynski BA, Lu Z. A 36-hospital time and motion study: how do medical-surgical nurses spend their time? *Perm J*. 2008; **12**(3): 25–34. Print.

Henning K, Ey S, Shaw D. Perfectionism, the impostor phenomenon and psychological adjustment in medical, dental, nursing and pharmacy students. *Med Educ*. 1998; **32**(5): 456–64. Print.

Hermans HJM, Gieser T. *Handbook of Dialogical Self Theory*. Cambridge: Cambridge University Press; 2011. Print.

Hermans HJM, Kempen HJG. *The Dialogical Self: meaning as movement*. San Diego: Academic; 1993. Print.

Heron J. A six-category intervention analysis. *Br J Guidance & Counselling*. 1976; **4**(2): 143–55. Print.

Herzberg F. *The Motivation to Work*. New York: Wiley; 1959. Print.

Hinduism Today. *Join the Hindu Renaissance*. Hinduism Today Magazine. Web. Available at: www.hinduismtoday.com (accessed 14 November 2011).

Hilbert D, Cohn-Vossen S. *Geometry and the Imagination*. 2nd ed. London: Chelsea Publishing Company; 1990. Print.

Hofstadter DR. *Gödel, Escher, Bach*. Harmondsworth: Penguin; 1980. Print.

Hume D. *A Treatise of Human Nature; being an attempt to introduce the experimental method of reasoning into moral subjects*. Cleveland: World Pub.; [1739] 1962. Print.

Hutton W. *The State We're In*. London: Jonathan Cape; 1995. Print.

Hymes J. editor. *The Child under Six*. London: Consortium; 1994. Print.

Ignatow D. *Against the Evidence: selected poems, 1934–1994*. [Middletown, Conn.]: Wesleyan University Press; 1993. Print.

Internet Encyclopedia of Philosophy. *Time*. Internet Encyclopedia of Philosophy. Web. Available at: www.iep.utm.edu/time (accessed 14 November 2011).

Isaksen SG, Treffinger DJ. *Creative Problem Solving: the basic course*. Buffalo, NY: Bearly; 1985. Print.

Isen A, Daubman KA, Nowicki GP. Positive affect facilitates creative problem solving. *J Pers Soc Psychol*. 1987; **52**(6): 1122–31. Print.

Ivancevich JM, Matteson MT. Stress and work: a managerial perspective. In: Quick JC, Bhagat RS, Dalton JE, Quick JD, editors. *Work Stress: health care systems in the workplace*. New York: Praeger; 1980. pp. 27–49. Print.

James W. *The Principles of Psychology*. Charleston, SC: BiblioLife; 2010. Print.

Juran JM, Gryna FM. *Juran's Quality Control Handbook*. New York: McGraw-Hill; 1988. Print.

Kabat-Zinn J. *Full Catastrophe Living: using the wisdom of your body and mind to face stress, pain, and illness*. New York, NY: Dell Pub., a Division of Bantam Doubleday Dell Pub. Group; 1991. Print.

Kahn RL, Byosiere P. Stress in organizations. In: Dunnette MD, Hough LM, editors. *Handbook of Industrial and Organizational Psychology, Vol. 3*. Palo Alto, CA: Consulting Psychologists Press; 1992. pp. 571–650. Print.

Kahneman D. *Thinking, Fast and Slow*. New York: Penguin; 2012. Print.

Kandel ER, Schwartz JM, Jessell TM. *Principles of Neural Science*. New York: McGraw-Hill, Health Professions Division; 2000. Print.

Kant I. *Groundwork for the Metaphysics of Morals*. New Haven: Yale University Press; 2002. Print.

Kaufman JC, Beghetto RA. Beyond big and little: the Four C Model of Creativity. *Rev Gen Psychology.* 2009; **13**: 1–12. Print.

Keating T. Centering Prayer. Web. Available at: www.centeringprayer.com (accessed 12 November 2011).

King LS. *Medical Thinking: a historical preface*. Princeton, NJ: Princeton University Press; 1982. Print.

Kleinke CL, Peterson TR, Rutledge TR. Effects of self-generated facial expressions on mood. *J Pers Soc Psychol.* 1998; **74**(1): 272–9. Print.

Kleinman A. *Patients and Healers in the Context of Culture: an exploration of the borderland between anthropology, medicine, and psychiatry*. Berkeley: University of California; 1980. Print.

Ko U. Ananda. *Beyond Self: 108 Korean Zen poems*. Berkeley, CA: Parallax; 1997. Print.

Koch R. *The Natural Laws of Business: applying the theories of Darwin, Einstein, and Newton to achieve business success*. New York: Currency/Doubleday; 2001. Print.

Koestler A. *The Ghost in the Machine.* London: Hutchinson; 1967. Print.

Kolb DA. *Experiential Learning: experience as the source of learning and development*. Englewood Cliffs, NJ: Prentice-Hall; 1984. Print.

Kornfield J. *Buddha's Little Instruction Book*. London: Rider & Co; 1996. Print.

Kotter JP. *Leading Change*. Boston, MA: Harvard Business School; 1996. Print.

Kumar M. *Quantum: Einstein, Bohr, and the great debate about the nature of reality*. New York: W.W. Norton; 2009. Print.

Kurtz SM, Silverman J, Draper J. *Teaching and Learning Communication Skills in Medicine*. Oxford: Radcliffe Publishing; 2005. Print.

Lalor D. *Creating a Therapeutic Environment. Counselling in Perth, Western Australia.* Cottesloe Counselling Centre. Web. Available at: www.cottesloecounselling.com.au (accessed 24 October 2011).

Lazarus RS, Folkman S. *Stress, Appraisal, and Coping*. New York: Springer; 1984.

Launer J. *Narrative-based Primary Care: a practical guide*. Oxford: Radcliffe Medical Press; 2002. Print.

Légaré F, Ratté S, Stacey D, *et al*. Interventions for improving the adoption of shared decision making by healthcare professionals. *Cochrane Database Syst Rev.* 2011; **10**: CD001431. Web.

Lehrer J. *Imagine: how creativity works*. Edinburgh: Canongate; 2012. Print.

Levensky E, Forcehimes A, Beitz K. Motivational interviewing: an evidence-based approach to counseling helps patients follow treatment recommendations. *Am J Nurs.* 2007; **107**(10): 50–8. Print.

Lewin S, Skea Z, Entwistle V, *et al*. Effects of interventions to promote a patient-centred approach in clinical consultations. *Cochrane Database Syst Rev.* 2001; **4**: CD00326. Web.

Lewin SA, Skea Z, Entwistle VA, *et al*. Interventions for providers to promote a patient-centred approach in clinical consultations. *Cochrane Database Syst Rev.* 2012; **12**: CD003267. Print.

Linehan M. *Cognitive Behavioural Treatment of Borderline Personality Disorder*. London: Guildford; 1993. Print.

Linn LS, Yager J, Cope D, Leake B. Health status, job satisfaction, job stress, and life satisfaction among academic and clinical faculty. *JAMA*. 1985; **254**(19): 2775–82. Print.

Living Life to the Full. *Free Online Skills Course*. Living Life to the Full. Web. Available at: www.llttf.com (accessed 28 October 2011).

Locke J, Bassett T, Holt E. *An Essay Concerning Humane Understanding: in four books*. London: Printed by Eliz. Holt for Thomas Basset; 1690. Print.

Mackenzie RA. *The Time Trap*. New York: AMACOM; 1972. Print.

Maslach C, Schaufeli W, Leiter M. Job burnout. *Annu Rev Psychol*. 2001; **52**: 397–422. Web.

Maslow AH. A theory of human motivation. *Psychol Rev.* 1943; **50**(4): 370–96. Print.

Maslow AH. *The Farther Reaches of Human Nature*. New York: Penguin; 1976. Print.

May R. *The Courage to Create*. London: Collins; 1976. Print.

McCambridge J. Motivational interviewing is equivalent to more intensive treatment, superior to placebo, and will be tested more widely. *Evidence-Based Mental Health*. 2004. **7**(2): 52. Print.

McKinlay JB, Potter DA, Feldman DA. Non-medical influences on medical decision-making. *Soc Sci Med*. 1996; **42**(5): 769–76. Print.

McQuaid JR, Carmona PE. *Peaceful Mind: using mindfulness and cognitive behavioral psychology to overcome depression*. Oakland, CA: New Harbinger; 2004. Print.

McVicar A. Workplace stress in nursing: a literature review. *J Adv Nurs*. 2003; **44**(6): 633–42. Print.

Melville A. Job satisfaction in general practice: implications for prescribing. *Soc Sci Med. Part A: Medical Psychology & Medical Sociology.* 1980; **14**(6): 495–9. Print.

Mitchley SE. The medical interview: the three-function approach. *Postgrad Med J*. 1992; **68**(799): 397–8. Print.

MoodGYM. Welcome. Web. Available at: www.moodgym.anu.edu.au (accessed 28 October 2011).

Moran P. *Antisocial Personality Disorder*. London: Gaskell; 1999. Print.

Morrison T. *Staff Supervision in Social Care: making a real difference for staff and service users*. Brighton: Pavilion; 2005. Print.

National Institute for Health and Care Excellence. *Anxiety: management of anxiety (panic disorder, with or without agoraphobia, and generalised anxiety disorder) in adults in primary, secondary and community care*. NICE. March 2011. Web. Available at: http://guidance.nice.org.uk/CG22

National Institute for Health and Care Excellence. *Brief Interventions and Referral for Smoking Cessation in Primary Care and Other Settings*. NICE. 2006. Web. Available at: www.nice.org.uk/nicemedia/pdf/SMOKING-ALS2_FINAL.pdf

National Institute for Health and Care Excellence. *Cognitive Behavioural Therapy for the Management of Common Mental Health Problems*. NICE. December 2010. Web. Available at: www.nice.org.uk/usingguidance/commissioningguides/cognitivebehavioural therapyservice/cbt.jsp

National Institute for Health and Care Excellence. *Computerised Cognitive Behaviour Therapy for Depression and Anxiety: review of Technology Appraisal 51*. NICE. February 2006. Web. Available at: www.nice.org.uk/nicemedia/pdf/TA097guidance.pdf

Neighbour R. *The Inner Consultation: how to develop an effective and intuitive consulting style*. Lancaster: MTP; 1987. Print.

NHS Centre for Reviews. *Effectiveness Matters: counselling in primary care*. 2001; **5**(2): n.p. Print.

NHS Direct. *Decision Aids*. NHS Direct. Web. Available at: www.nhsdirect.nhs.uk/decisionaids.

NHS Institute for Innovation and Improvement. *Releasing Time to Care: the productive ward*. 2007. Available at: www.institute.nhs.uk/quality_and_value/productivity_series/productive_ward.html.

Noonuccal, Oodgeroo. *My People*. 3rd ed. Milton, QA: The Jacaranda Press; 1990. Print.

Ogedegbe G. Labeling and hypertension: it is time to intervene on its negative consequences. *Hypertension*. 2010; **56**(3): 344–5. Print.

O'Hara LA. Creativity and intelligence. In: Sternberg RJ, editor. *Handbook of Creativity*. Cambridge University Press; 1999. Print.

Open Door Coaching. *Job Satisfaction Inventory*. Open Door Coaching. Web. Available at: www.opendoorcoaching.com/PDF%20files/Job%20Satisfaction%20Inventory.PDF (accessed 24 October 2011).

Orwell G. *Nineteen Eighty-four, a novel*. New York: Harcourt, Brace; 1949. Print.

'Overcoming' series. Constable & Robinson Publishers. Web. Available at: www.overcoming.co.uk

Paice E, Moss F. How important are role models in making good doctors. *BMJ*. 2002; **325**: 707. Print.

Patient.co.uk. *Significant Event Analysis*. Health Information and Advice, Medicines Guide, Patient.co.uk. Web. Available at: http://patient.co.uk (accessed 24 October 2011).

Patrick CJ, Craig KD, Prkachin KM. Observer judgments of acute pain: facial action determinants. *J Pers Soc Psych*. 1986; **50**(6): 1291–8. Print.

Pendleton D, Schofield T, Tate P, Havelock P. *The Consultation: an approach to learning and teaching*. Oxford: Oxford University Press; 1984. Print.

Penrose R. *The Emperor's New Mind: concerning computers, minds, and the laws of physics*. Oxford: Oxford University Press; 1989. Print.

Pepler D J. Play and divergent thinking. In: Pepler DJ, Rubin KH. *The Play of Children: current theory and research*. Basel; New York: Karger; 1982. Print.

Pepler DJ, Rubin KH, editors. *The Play of Children: current theory and research*. Basel; New York: Karger; 1982. Print.

Prkachin KM. Dissociating spontaneous and deliberate expressions of pain: signal detection analyses. *Pain*. 1992; **51**(1): 57–65. Print.

Prochaska JO, DiClemente CC. *The Transtheoretical Approach: crossing traditional boundaries of therapy*. Malabar, Florida: R. E. Krieger; 1994. Print.

Proshansky H. The field of environmental psychology. *Handbook of Environmental Psychology*. New York: Wiley; 1987. Print.

Proshansky H, Fabian A, Kaminoff R. Place-identity: physical world socialization of the self. *J Environ Psychol*. 1983; **3**(1): 57–83. Print.

Quakers. *Quaker Faith & Practice: the book of Christian discipline of the yearly meeting of the Religious Society of Friends (Quakers) in Britain*. London: Yearly Meeting of the Religious Society of Friends (Quakers) in Britain; 2009. Print.

Reuler JB, Nardone DA. Role modeling in medical education. *West J Med*. 1994; **160**(4): 335–7. Print.

Rolfe G, Freshwater D, Jasper M. *Critical Reflection for Nursing and the Helping Professions: a user's guide*. Houndmills, Basingstoke, Hampshire: Palgrave; 2001. Print.

Rossman J. *Industrial Creativity; the psychology of the inventor*. New Hyde Park, NY: University; 1964. Print.

Roter DL, Frankel RM, Hall JA, Sluyter D. The expression of emotion through nonverbal behavior in medical visits. Mechanisms and outcomes. *J Gen Intern Med*. 2006; **21**(Suppl. 1): S28–34. Print.

Sackett DL, Rosenberg WM, Gray JA, *et al*. Evidence based medicine: what it is and what it isn't. *BMJ*. 1996; **312**: 71–2. Print.

Sandman, L, Munthe C. Shared decision making, paternalism and patient choice. *Health Care Anal*. 2010; **18**(1): 60–84. Print.

Schegloff EA, Jefferson G, Sacks H. The preference for self-correction in the organization of repair in conversation. *Language*. 1977; **53**: 361–82. Print.

Schön DA. *The Reflective Practitioner: how professionals think in action*. Aldershot: Ashgate; [1983] 2002. Print.

Schwarz, B. *The Paradox of Choice: why more is less*. HarperCollins; New edition; 2005. Print.

Searle JR. *Mind: a brief introduction*. Oxford: Oxford University Press; 2004. Print.

Segal Z, Williams JM, Teasdale J. *Mindfulness-Based Cognitive Therapy for Depression: a new approach to preventing relapse*. New York: Guildford; 2001. Print.

Seligman MEP. *Authentic Happiness: using the new positive psychology to realize your potential for lasting fulfillment*. New York: Free; 2002. Print.

Sharot T, De Martino B, Dolan RJ. Neural activity predicts attitude change in cognitive dissonance. *Nature Neuroscience*. 2009; **29**(12): 3760–5. Print.

Silverman J, Kurtz SM, Draper J. *Skills for Communicating with Patients*. 3rd ed. London: Radcliffe Publishing; 2013. Print.

Simon HA. The mind's eye in chess. In: Chase WG, editor. *Visual Information Processing*. New York: Academic; 1973. Print.

Simon P, Garfunkel A. *The Sounds of Silence*. Columbia, released 1965. CD.

Simonton DK. Creativity, leadership, and chance. In: Sternberg RJ, editor. *The Nature of Creativity*. Cambridge: Cambridge University Press; 1988. Print.

Smith HW. *The 10 Natural Laws of Successful Time and Life Management: proven strategies for increased productivity and inner peace*. New York, NY: Warner; 2003. Print.

Snyder CR, Lopez SJ, editors. *Handbook of Positive Psychology*. Oxford: Oxford University Press; 2009. Print.

Soria R, Legido A, Escolano C. A randomised controlled trial of motivational interviewing for smoking cessation. *Br J Gen Pract*. 2006; **1**(56): 531. Print.

Sowa JF. 'Representing knowledge soup in language and logic'. Available online at: www.jfsowa.com/talks/souprepr.htm

Sternberg RJ. *Beyond IQ: A Triarchic Theory of Intelligence*. Cambridge: Cambridge University Press; 1985.

Stewart I, Joines V. *TA Today: a new introduction to transactional analysis*. Nottingham: Lifespace Pub.; 1987. Print.

Stewart M, Roter D. *Communicating with Medical Patients*. Newbury Park: Sage Publications; 1989. Print.

Stiglitz JE, Sen A, Fitoussi J-P. *Report by the Commission on the Measurement of Economic Performance and Social Progress*. Paris: Commission; 2009. Print.

Stott NC, Davis RH. The exceptional potential in each primary care consultation. *J R Coll Gen Pract*. 1979; **29**: 201–5. Print.

Suzuki DT. *Essays in Zen Buddhism, third series*. London: Published for the Buddhist Society by Rider; 1958. Print.

Suzuki S, Dixon T. *Zen Mind, Beginner's Mind*. New York: Walker/Weatherhill; 1970. Print.

Tarski A. *Logic, Semantics, Metamathematics; papers from 1923 to 1938*. Oxford: Clarendon; 1956. Print.

Taylor D, Bury M. Chronic illness, expert patients and care transition. *Sociology of Health & Illness*. 2007; **29**(1): 27–45. Print.

Tellegen A, Lykken DT, Bouchard TJ, *et al*. Personality similarity in twins reared apart and together. *J Pers Soc Psychol*. 1988; **54**(6): 1031–9. Print.

Thich Nhat Hanh, Mobi Ho, Vo-Dinh Mai. *Miracle of Mindfulness: an introduction*. Boston: Beacon; 1975. Print.

Top Nursing Colleges. *Nursing Theories and Sub-theories*. Top Nursing Colleges. Web. Available at: www.topnursingcolleges.com/nur/nursing-theories-and-sub-theories.html (accessed 12 November 2011).

Tsao L. How much do we know about the importance of play in child development. *Childhood Educ*. Summer 2002. Findarticles.com. Web. Available at: http://findarticles.com/p/articles/mi_qa3614/is_200207/ai_n9147500

Tuckett D, Boulton M, Olson C, Williams A. *Meetings between Experts: an approach to sharing ideas in medical consultations*. London: Tavistock, 1985. Print.

Ubel PA, Angott AM, Zikmund-Fischer BJ. Physicians recommend different treatment for patients than they would choose for themselves. *Arch Intern Med*. 2011; **171**(18): 630–4. Print.

Ulrich RS. How design impacts wellness. *Healthc Forum J.* 1992; **35**(5): 20–5. Print.

Upton J. *Comments*. FearFighter for Panic and Anxiety. Web. Available at: www.fearfighter.com (accessed 28 October 2011).

US National Cancer Institute. *Cancer Screening Overview (PDQ®)*. US National Cancer Institute. Web. Available at: www.cancer.gov/cancertopics/pdq/screening/overview/HealthProfessional/page1 (accessed 24 October 2011).

Van Ham I, Verhoeven A, Groenier K, Groothoff J and De Haan J. Job satisfaction among general practitioners: A systematic literature review. *Eur J Gen Pract.* 2006, **12**(4): 174–80. (doi:10.1080/13814780600994376)

Van Veen V, Krug MK, Scooler JW, Carter CS. Neural activity predicts attitude change in cognitive dissonance. *Nature Neuroscience.* 2009; **12**(11): 1469–74. Print.

Vandervert L, Schimpf P, Liu H. How working memory and the cerebellum collaborate to produce creativity and innovation. *Creativity Res J.* 2007; **19**(1): 1–18. Print.

Various. Evidence based practice in clinical hypnosis. *IJCEH.* 2007; **55**(2): n.p. Print.

Walker L. *Consulting with NLP: Neuro-linguistic Programming in the medical consultation*. Oxford: Radcliffe Medical Press; 2002. Print.

Wallas G. *The Art of Thought*. New York: Harcourt, Brace; 1926. Print.

Warren KS. *Coping with the Biomedical Literature: a primer for the scientist and the clinician*. New York, NY: Praeger; 1981. Print.

Waskett C. An integrated approach to introducing and maintaining supervision: the 4S Model. *Nurs Times.* 2009; **105**(17): 24–6. Print.

Weisberg RW. *Creativity: beyond the myth of genius*. New York: W.H. Freeman; 1993. Print.

West C. Against our will: male interruptions of females in cross-sex conversation. *Annals of the New York Academy of Sciences.* 1979 (Language, Sex); **327**(1): 81–96. Print.

White M. *Maps of Narrative Practice*. New York: W.W. Norton & Co; 2007. Print.

White M, Epston D. *Narrative Means to Therapeutic Ends*. New York: Norton; 1990. Print.

Wilber K. *A Brief History of Everything*. Boston, MA: Shambhala; 2007. Print.

Wilber K. An integral theory of consciousness. *J Consciousness Stud.* 1997; **4**(1): 71–92. Print.

Williams CJ, Garland A. Cognitive-behavioural therapy assessment model for use in clinical practice. *Adv Psych Treat.* 2002; **8**: 172–79. Print.

Williams ES, Konrad TR. Physician, practice, and patient characteristics related to primary care physician physical and mental health: results from the Physician Worklife Study. *Health Services Res.* 2002; **37**(1): 119–41. Print.

Williams ES, Konrad TR, Scheckler WE, *et al.* Understanding physicians' intentions to withdraw from practice: the role of job satisfaction, job stress, mental and physical health. *Health Care Manage Rev.* 2010; **35**(2): 105–15. Web.

Wilson PM, Kendall S, Brooks F. The Expert Patients Programme: a paradox of patient empowerment and medical dominance. *Health & Social Care in the Community.* 2007; **15**(5): 426–38. Web.

Yovel G, Kanwisher N. Face perception: domain specific, not process specific. *Neuron.* 2004; **44**(5): 889–98. Print.

Zhong E, Kenward K, Sheets V, *et al.* Probation and recidivism: remediation among disciplined nurses in six states. *Am J Nurs.* 2009; **109**(3): 48–57. Print.

CPD with Radcliffe

You can now use a selection of our books to achieve CPD (Continuing Professional Development) points through directed reading.

We provide a free online form and downloadable certificate for your appraisal portfolio. Look for the CPD logo and register with us at: www.radcliffehealth.com/cpd